Everything in life from marriage to manslaughter turns on the speed and cost at which men, things and thoughts can be shifted from one place to another.

Rudyard Kipling

The age of Machine Power on the farm, as well as in the cities, is upon us. It is the latest advanced step civilisation has taken. It is full of the promise to lift the great burden of drudgery, and wearisome, wearing out toil, from the muscles of man and carrying them, with much more profit, by power of Machinery, with the result that larger production, more profits and greater wealth will come, with the accompanying larger comforts, benefits and progress to humanity everywhere.

From *The Fordson at Work*, published by the Ford Motor Company, Highland Park, Michigan, USA in 1921

Fordson
Major

FORD
TRACTOR CONVERSIONS

STUART GIBBARD

Old Pond Publishing

First published 1995
Reprinted 2003

ISBN 1 903366 42 9

A catalogue record for this book is available from the British Library

Published by
Old Pond Publishing
104 Valley Road, Ipswich IP1 4PA
United Kingdom

www.oldpond.com

Distributed in North America by Diamond Farm Enterprises,
Box 537, Alexandria Bay, NY 13607, USA

Front cover photograph: County Super Four
Back cover photographs, top: Doe Triple D, 1962; bottom: Muir-Hill 121 Series 3

Cover design and book layout by Liz Whatling
Printed and bound in Great Britain by Butler & Tanner Ltd, Frome & London

Contents

Acknowledgements

I owe the following people and companies my sincere gratitude, as without their help, this book would not have been possible.

For help with information on County, I would like to thank Geoffrey Tapp, David Henderson (County Tractors Ltd.), Stan Anderson, Colin Bennett and Bruce Keech. I am grateful for the Doe story as provided by Alan E. Doe (Ernest Doe & Sons Ltd.), George Pryor, Eileen and Derek Hockley. Information on Chaseside and Northrop was kindly given by Gilbert McIntosh, Les Carter, Barry Booth and Jennie Starkey (JCB Archives). Kind assistance with Bray was given by Eddie Stevens and John Suckling, and the Matbro story could never have been written without the help of John Mathew, with additional information from Geoff Matthews, Mr A. J. Clarke, David Barnard and Geoff Ashcroft. Last, but by no means least, I would like to thank Ernest Briggs, David Lloyd (Lloyd Loaders), David J. B. Brown and Miss Vanna Skelley (University of Glasgow Business Records Centre) for help with the Muir-Hill story.

Special thanks go to the following five individuals who have who have provided considerable assistance, general information, contacts and photographs: David Bate, Richard Cavill, Ven Dodge and David Woods, and Jan Wiggers in Holland.

I am especially grateful to John Blackbeard for the loan of many *Arable Farming* photographs, and to David Cousins, features editor of *Farmers Weekly* (Reed Business Publishing), for permission to use cutaway drawings from *Power Farming, Farm Mechanization* and *British Farmer and Stockbreeder.* I am also greatly indebted to New Holland Ford for help received from various departments.

I would like to thank Rob Gibbard for reading my manuscript – he would not know a Ford tractor conversion if he fell over it in a field, but he can spot a typographical error a mile off–and my wife, Sue, for her patience, help and advice.

I am also very grateful to all the following people who have lent photographs, allowed permission for photographs to be used, provided information or helped in some way – Aardvark Clear Mine Ltd., Mike Alsop (Dowdeswell Engineering), Babcock International Group plc, James Baldwin, Arthur Battelle, Brian Bell, Edward Bourn, Stephen Burtt, James Cochrane, David Collings, Mark Farmer, Bob Friedlander (New Holland Ford), Nigel Ford, Rodney Gibson, Roger Haines (Agrosave), Ken Hatter, Graham Holmes (Bovington Tank Museum), Paul Hoptroff (New Holland Ford), Arthur Ingram, Bill King, Edgar Lancaster, Tom Lowther, Barry Milsom, Andrew Offer, Mark Osborne (A. T. Osborne Ltd.), Silsoe Research, A. N. Rogers, Eric Sixsmith, Graham Smitheringale, Mervyn Spokes, Johnny Weal, Billy Wilson (Jas. P. Wilson).

Author's Note

All horsepower figures used in the book are approximate and are included only for informal comparison. Where possible, manufacturers' quoted figures conforming to British Standard (BS) ratings are given. However, most tractors built, or sourced from, the United States are rated in SAE (Society of Automobile Engineers, USA) figures.

Further anomalies arise with the use of industrial engines. Ford often gave 'bare' engine test figures, which varied from an 'installed' engine test. Several manufacturers also quoted gross (overload) horsepower figures as opposed to 'continuous' output.

All weights and measures are given in imperial figures, as this was the system in common usage when most of the machines featured in the book were built.

Preface

This book was conceived from a desire to draw together the comparative histories of County, Doe, Chaseside, Muir-Hill, Matbro and Bray in one volume.These well known British companies were all leaders in their own fields of engineering design and innovation - each producing successful ranges of agricultural, earthmoving or construction machinery.

Pioneering use of four-wheel drive in Britain further linked these trail-blazing concerns. During the 1950s and 1960s, each company, using its own particular and unique systems, developed and built four-wheel drive agricultural tractors, usually based on a Ford skid unit.

And it was this use of Ford and Fordson tractors as basic units on which to develop their machines that provides the common denominator between these firms, hence the book title, *Ford Tractor Conversions*.

But the title still does not fit the content perfectly. Most of the companies did not, by a long way, base all their machines on Ford skid units - and I make no apologies for including other makes of tractor. Furthermore, 'conversions' is not a word that does adequate justice to the specialised equipment and machinery built by these concerns.

I have tried to put the work of these companies into context by showing how and why the use of the Ford skid unit developed, and illustrate some of the other conversions built on it. The list is by no means complete, as to mention all the hundreds of conversions built worldwide would be an impossible task.

Many readers will be surprised that Roadless appears to have only a brief mention in the book. Roadless were indeed pioneer builders of four-wheel drive conversions and extensive users of Ford skid units, but much of their important early work was built around so many different makes and types of vehicle that it could never be properly covered in the confines of a book on Ford tractor conversions, and justifies a separate book of its own.

During the eighteen months of research for *Ford Tractor Conversions*, I have met and spoken with many current and former directors and employees of the companies concerned, who have all readily shared their knowledge and memories (and sometimes, even their lunch) with me. These are marvellous and genuine people whose passion and devotion to the work, and loyalty to the companies are undiminished. I can only hope that this book conveys at least a small part of their enthusiasm to the reader.

I have also received tremendous help from many companies still involved today, and have been given access to important company records and photographs, gaining an unparalleled insight into their work and developments, for which I am very grateful.

This is the story of magnificent men working for proud companies that deserve their place in history, and I have enjoyed every minute of writing it.

Stuart Gibbard
May 1995

CHAPTER 1

Ford Tractor Conversions

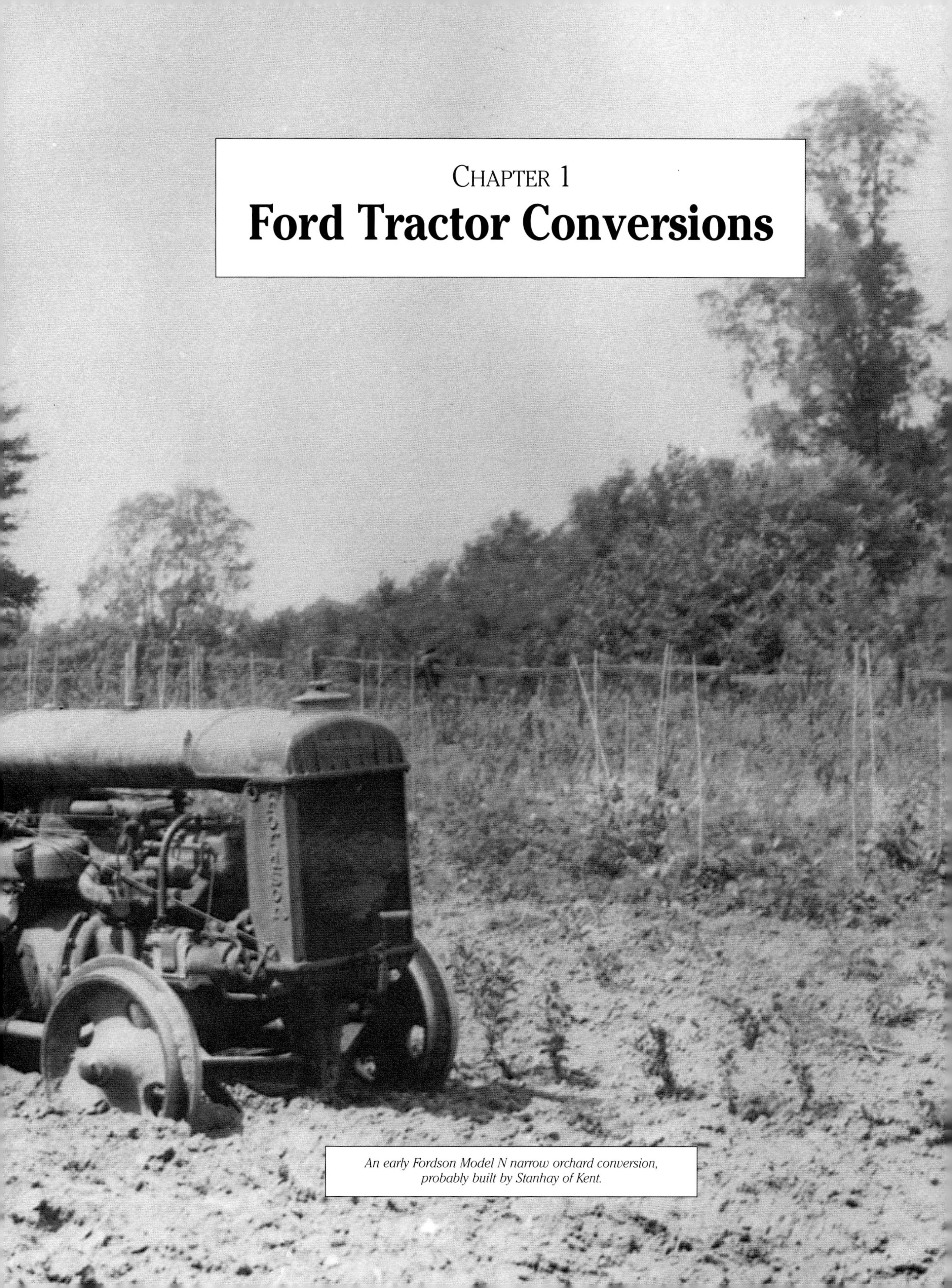

An early Fordson Model N narrow orchard conversion, probably built by Stanhay of Kent.

Without a doubt, the Ford has probably been the subject of more conversions than any other make of tractor. Even Henry Ford's first experimental tractors were conversions themselves, built from various automobile components.

In Britain, few manufacturers of industrial and construction machines have not used a Ford skid unit or power plant at some time, and most firms pioneering four-wheel drive conversions started with Fordson tractors. Even today, all over the world, specialist machines, from orange and lemon tree shapers in Israel to minefield clearance vehicles in the Gulf, are based on Ford tractor skid units. In the early days, the Ford was such a popular choice for conversion because, put simply, it was there!

Any manufacturer engaged in the small-scale production of plant or machinery could not afford to develop his own engines and transmissions, so the obvious answer was to use the Fordson as a skid unit. And the Fordson was available when little else was. During the 1920s and 1930s there were very few British tractor manufacturers in existence, and of those that were, none could match the Fordson for price, simplicity or mass production in numbers. The thought of Muir-Hill choosing something like the cumbersome Birmingham-built Alldays and Onions tractor, weighing over three tons and costing £630, on which to base their dumpers in the early 1920s, is unimaginable, when the compact 23 cwt Fordson F could be bought for £120.

The first Model F Fordson left the Brady Street and Michigan Avenue plant in Dearborn on 8 October 1917, and during part of the 1920s, it accounted for 75 per cent of tractor production in the United States. The introduction of the Fordson was seen as an opportunity to bring cheap machine power to the masses, not just to the farmer, but also in factories and industry. The Fordson tractor was an economical and lightweight power source, and the rigid unit construction of its engine and transmission, requiring no supporting frames, made it the ideal basis for conversion into other machines. The

BELOW: *An Aveling and Porter steam-roller towers over an early road-roller conversion of the Fordson Model F.*

publicity booklet for the tractor, *The Fordson at Work*, published in the USA by the Ford Motor Company in 1921, listed over one hundred uses for the Fordson, including driving a Ferris wheel, washing machine operation, and as an industrial railway locomotive. When production of the Fordson was discontinued in the USA in 1928, a total of 739,978 tractors had been built.

The Fordson came to Britain first from Dearborn, and then from a factory in Cork, Ireland, where it evolved into the Model N in 1929. Many of the early conversions on Ford tractors in Britain were born out of attempts by dealers to improve sales in the depressed 1920s by building specialist machines

ABOVE: An interesting military use of the Fordson F skid unit to power the Newton Tractor. Designed by Colonel Newton as a simple tracked load carrier for use in the First World War, it was intended for mass production in the USA by Buick and Studebaker, but never went beyond the prototype stage.

LEFT: A Model F Fordson, fitted with Trackson tracks, powers a Canadian Sawyer-Massey one-man power maintainer clearing snow from Quebec city streets in 1926.

ABOVE:
A Fordson fitted with American Trackson tracks and a snow plough blade

for industrial uses. Tractors that the farmers were not buying were converted into dumpers, cranes, loading shovels and shunting tractors for the construction industry and building trade, creating new markets for the Fordson in the process.

The new Dagenham plant in Essex came on-line in 1931, and tractor production was transferred there at the end of the following year, rising steadily as demand increased. The Ford Motor Company also set up the Industrial Unit Department at Dagenham, to stimulate sales of power plants and skid units to specialist manufacturers and industrial users.

The Fordson tractor skid unit was only one of the products promoted by this new department which also sold industrial versions of the 8 hp Model Y, 14.9 hp Model BF and 24 hp Model B Ford car and commercial vehicle engines. These engines were used by many engineering firms

RIGHT:
The Malcolm Roller, an Australian conversion of the Model F into an eight ton road-roller, built by Malcolm Moore Limited of Melbourne.

ABOVE:
Pattisson Super Beta Golf Tractor, based on the Ford A engine and chassis, built in Stanmore, Middlesex.

LEFT:
One of many industrial Fordson tractors that found work with municipal coporations throughtout Britain. This 1930s model, fitted with a coachbuilt cab, is seen towing an Eagle trailer.

ABOVE: Pattisson Super Beta Lightweight Tractor, designed for golf course and playing field maintenance, based on the Ford one-ton chassis powered by the 24 hp Model AE engine.

RIGHT: Another golf tractor based on the Ford A chassis and engine and built by Allan Taylor of Wandsworth.

and were incorporated into a variety of equipment, including compressors, welders, fire pumps and rail cars. Marine versions of the engines were also produced. Many of these industrial engines were eventually used in agriculture, powering farm machinery and some smaller tractors, especially after the 30 hp V8 was introduced.

The Industrial Unit Department also supplied some complete car or commercial vehicle chassis units with engines and transmissions, for conversion into purpose-built tractors for golf course and playing field maintenance. Before long, the Ford Motor Company was actively encouraging specialist manufacturers to use the Fordson as a skid unit for their machines, as well as further promoting the tractor for use in industry, building sites and municipal service, both as a prime-mover and a stationary power plant.

With the outbreak of the Second World War, demand for the Fordson exceeded all expectations. During the war, the company tried to maintain production at one hundred tractors a day, with the exception of the times when bomb damage interfered with manufacture. By the end of the war, 120,000 Fordsons had been sold on the domestic market, and this figure did not include export sales. The thousands of Fordsons that went on the farms and helped put over five million extra acres of arable land into production in England and Wales between 1939 and 1944 are only part of the story.

The Admiralty, the War Office and the Royal Air Force all required machines for innumerable tasks. Muir-Hill dumpers were produced for the Forces, and Roadless half-tracks equipped with winches were used extensively on the airfields during the Battle of Britain. Refinements were cast aside for the duration, and basic machines, such as Aveling Barford's Utility four-ton road roller, needed to be produced as quickly and cheaply as possible, using a readily available skid unit. As 95 per cent of all the tractors built in Britain during the war came from Dagenham, it was only natural that the Fordson would be the basis of all these machines.

In 1943, Ford were approached by the Ministry of Agriculture to design a new tractor more suited to rowcrop work. After several prototypes and much testing, the first Fordson E27N Major rolled off the Dagenham assembly line in March 1945. Wartime limitations had not allowed for the development of a new engine, but the E27N had an improved single plate clutch, and the old worm final-drive had been finally replaced by a new rear transmission incorporating a crown wheel and pinion.

In the aftermath of war, Britain's blitzed towns and cities required much extensive reconstruction work – bombsites had to be cleared and houses and factories needed rebuilding. In 1946, the Labour government passed the New Towns Act, authorising further development and outlining plans for the creation of up to twenty new towns. Construction machinery manufacturers were faced with a post-war boom in sales for their products, and the E27N Major proved a reliable and economical source of power for their machines.

As the E27N Major became more popular for conversion, the old Fordson side-valve engine design began to show its age. Still

LEFT:
The Newbury self-propelled potato harvester, powered by the Ford V8 engine and manufactured by the Victory Agricultural Machine Company of Berkshire.

BELOW:
The experimental Leamington half-track conversion of the Fordson tractor. Built in Ford's own foundry at Leamington, it appears similar to the designs of A. J. Hosier, but uses Bren-gun carrier parts and tracks.

RIGHT: A 1935 winch tractor, based on the Fordson N, built by the Automower Engineering Company of Bath.

BELOW: An Aveling Barford wartime Utility four ton roller, based on the Fordson.

basically unchanged since its inception in 1917, the TVO engine was underpowered at 29 hp for many of its new industrial uses. Also, its outdated gravity-fed lubrication system led to some oil starvation problems occurring when machines were operated on severe gradients. The Perkins P6 diesel engine option for the E27N was eagerly awaited, and was quickly adopted by the conversion builders soon after its introduction at the end of the 1940s.

The 200,000th Fordson E27N Major was completed on 29 January 1951, and was sold to New Zealand. Later the same year, 1,000 tractors went out to the Nile Delta in the Sudan. Both home sales and exports of the Fordson were on the increase, and another side to the conversion business was emerging. As the Fordson gained in popularity, both agriculture and the export markets demanded special variations, including crawlers, machines for sugar cane work, rowcrop and high clearance tractors. The Ford Motor Company were not geared up for what would be small production runs, and it was left to the conversion builders to provide these specialist machines.

When the new and greatly improved E1A Major was launched to the trade at the Kursaal, Southend-on-Sea, on Wednesday 21 November 1951, several representatives of the specialist builders, including Percy Tapp from County, Lt.-Colonel Philip Johnson from Roadless, and Edward Boydell from Muir-Hill, were on hand with details of their conversions for the new tractor.

Of most interest to the converters was the diesel version of the new overhead-valve

ABOVE: Another Aveling Barford product - a trench digger mounted on a Fordson N tractor.

ABOVE: An experimental wartime industrial Fordson fitted with a Brockhouse torque converter transmission for use as a towing tractor with the RAF.

engines. This proved to be an excellent power-unit, and before too long, the Fordson Diesel Major accounted for 99 per cent of total production.

Further export markets opened up, and the first shipment of Fordson Majors was made to the USA in September of 1953. The 500,000th Fordson to be built at Dagenham rolled off the assembly line on 22 February 1954, and in the years that followed, the new and uprated Power Major and subsequent Super Major models were introduced, along with the smaller Dexta.

The specialist manufacturers were kept busy supplying Fordson Major conversions to meet the increasing demands of the agricultural and industrial user, and the ever-widening export market. The call for agricultural tractors with increased traction was met by County, Roadless, Doe, Matbro and Bray, all of which introduced four-wheel drive machines based on the Fordson skid.

These machines could also benefit from more power after Ford introduced the 6D Diesel Industrial Engine in 1958. Designated the 590E, this six-cylinder engine was initially rated at 86 bhp, and as it was designed to bolt straight onto the Major transmission, it became widely used by the conversion builders. The 592E, an industrial version of the four-cylinder Fordson Major diesel engine, was also available, developing 55 bhp. The company still built a number of industrial petrol engines for other applications.

LEFT:
A 1949 Fordson E27N Major fitted with Straussler Lypsoid low ground-pressure tyres made by Nicholas Straussler Company of London.

The tractor skid units and industrial engines were all sold through the Industrial Sales Department, which operated from Ilford in Essex and in 1961 was supplying tractor skid units to twenty-seven major conversion builders in the UK. The skidded tractor unit was available in its minimum form as an engine, clutch and gearbox, and Ford defined the maximum build as 'a tractor which is complete except for some item which renders it incapable of being driven off the line. This is usually interpreted as being a pair of wheels.' Within these limits, virtually any build could be created to a customer's own

BELOW:
An experimental combine built by the National Institute of Agricultural Engineering (NIAE) at Silsoe in 1956, based on a Perkins P6 powered Fordson E27N fitted with a header built from Allis Chalmers components.

ABOVE:
A winch tractor built by Automower Engineering. Based on the Fordson Super Major, it is still used for electric pylon cable work.

LEFT:
A close-up of the Super Major power unit in the Automower machine. The machine is only two-wheel drive, with the heavy-duty front axle carrying the weight of the winch.

specifications, including the addition of a heavy-duty clutch or rear axle. The finished unit was supplied on a wooden skid, hence the name 'skid unit'.

In March 1961, Ford Tractor Operations was established by the American parent Ford Motor Company to coordinate the activities of the Tractor and Implements Division in the USA and the tractor production from Dagenham. This was the first step towards

ABOVE:
An Australian Hewco Twin Powered Scraper based on two Fordson Power Major units. The original engines have been replaced by 80 hp 2712E power plants.

LEFT:
A Fordson Major skid unit powers this soil borer mounted on a Ford Thames four-wheel drive truck built by AWD of Camberley.

RIGHT:
The prototype Irish Armer sugar beet harvester on demonstration, built around an early TVO version of the Fordson E1A Major.

BELOW:
The Epping Auto-Shunter built by F.E. Weatherill Limited, based on a Fordson Diesel Major.

creating the concept of an integrated world-wide range of Ford tractors.

A one hundred acre site was acquired at Basildon in 1961, and work started on a new tractor factory. This complex would not only accommodate tractor production from Dagenham, but would also house Product Engineering from the Rainham Works and Mudlands Farm, Sales and Marketing from Ilford and tractor parts distribution from Aveley. Plans were put in motion for the new 6X models, a standardised range of virtually identical tractors to be produced from different Ford factories for world-wide distribution.

The 6X range was launched at Basildon in

ABOVE:
A 1965 Ford 4000 skid unit fitted to a John Salmon self-propelled beet harvester.

LEFT:
An industrial Ford 5000 fitted with a Wickham Poole semi-trailer built for Taylor Woodrow.

ABOVE: A Ford 4610 skid unit is used to power a self-propelled Chafer sprayer.

1964. The four new models consisted of the three-cylinder 2000, 3000 and 4000 tractors, and the top of the range four-cylinder 5000. With the exception of the 2000, all the tractors were available with eight-speed manual gearboxes or the optional Select-O-Speed ten-speed transmission.

The new range was a mixed blessing to the conversion builders. All the conversions had to be totally redesigned to fit the new skid units, involving a lot of time and expense. However, the much-improved design and

RIGHT: A unique use of the Ford 8210 skid unit in the Aardvark mine clearance vehicle. Built in Scotland for military use, this machine uses a revolving flail to clear a path through minefields.

LEFT:
One of two Ford 7710 tractors, fitted with Atlas 3006 cranes, used in the construction of the Channel Tunnel.

features of the Ford 5000, which was the model most often specified as a skidded unit, allowed the specialist manufacturers to build far more sophisticated machines. Ford's policy of marketing identical tractor models across the world meant that the products exported by these smaller specialist companies would benefit from an established global service and parts network, leading to enhanced overseas sales.

The 3000, 4000 and 5000 tractors were all available as skidded units from the Tractor Operations Industrial Power Unit Department at Basildon. The Select-O-Speed transmission, an epicyclic gearbox allowing clutch-less gear changes, came in for some

BELOW:
An experimental forestry forwarder built by A T Osborne Limited of Romsey, Hampshire. Fitted with a driven rear axle, it is based on the Ford 7710 skid unit.

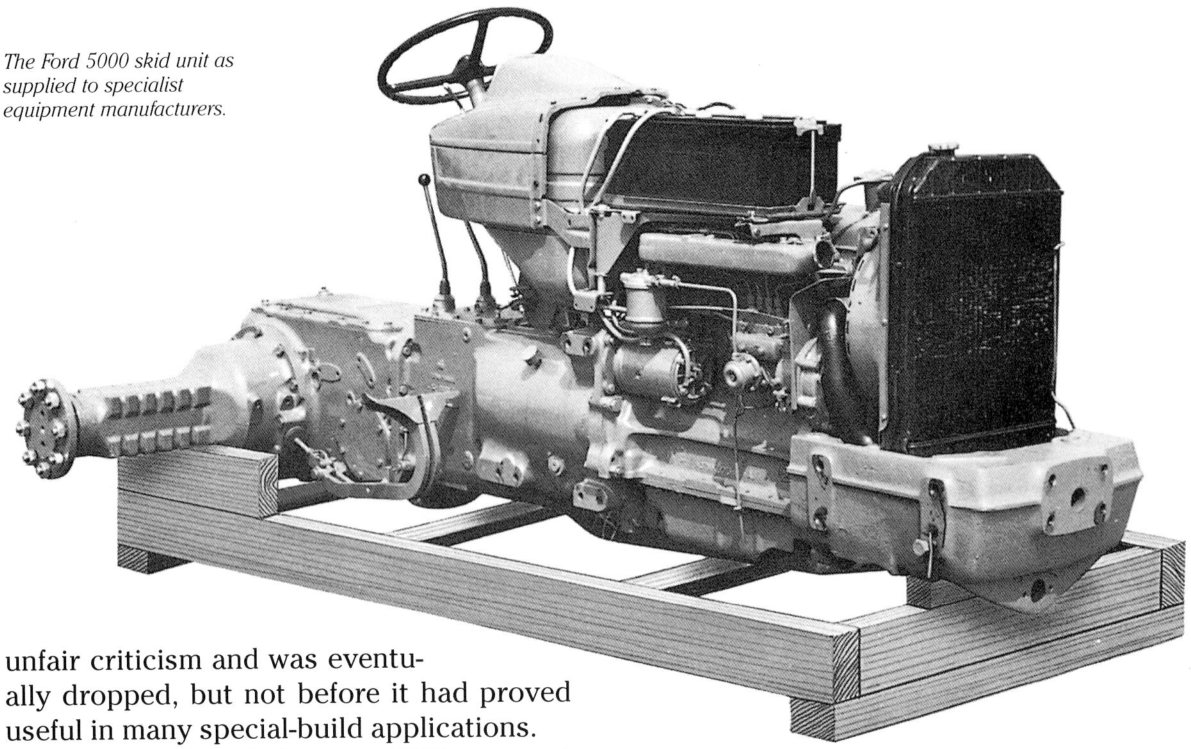

The Ford 5000 skid unit as supplied to specialist equipment manufacturers.

unfair criticism and was eventually dropped, but not before it had proved useful in many special-build applications.

Ford introduced the new 2700 range of industrial engines in 1965. Two vertical six-cylinder diesel engines were now offered, the 2703E, nominally rated at 89.5 bhp, and the more powerful (98 bhp) 2704E. An industrial version of the 2701E Ford 5000 engine was also available.

In 1971, Ford brought out the turbocharged 7000 tractor which was used as the basis of both the County 944 and the Roadless 94T four-wheel drive machines. The industrial engine range was revised the same year, and the upgraded 2713E, 2714E and 2715E six-cylinder units were introduced, offering even more horsepower.

As Ford brought out new tractor models, so the choice of skid units, engines and transmissions grew, giving the specialist equipment manufacturers and conversion builders the opportunity to increase their ranges and continually improve the specification of their machines. By the time the 10 Series was introduced in 1981, Ford was

RIGHT: A special-build Ford 5000 with a Select-O-Speed transmission, incorporated in the Century 125 tower crane built by Century Fabrications of Horsham, Sussex.

offering its own four-wheel drive versions of nearly all its tractors, a move which, while increasing the number of options available to the skidded unit customer, was to seriously damage sales for the manufacturers of four-wheel drive tractors and conversions, including County, Roadless, Selene and Schindler.

Skid unit sales were now handled through the Original Equipment Manufacturer (OEM) department at Basildon. Apart from handling the skid units, OEM also supplied individual power-train components, including engines, transmissions and drive axles.

Following the acquisition of New Holland and Versatile in 1987, Ford Tractor Operations was restructured into a new organisation called New Holland Ford, with a product range of tractors from 8 to 470 hp. The OEM department now offered the specialist equipment manufacturers a choice of ten diesel engines, seven transmissions including torque converter models, several drive and drive-steer axles and fifteen complete power-trains. The OEM customers included agricultural, forestry, mining, industrial and materials handling equipment manufacturers, who had the advantage of Ford's network of over 5,000 dealers providing parts and service support in more than 130 countries of the world.

In 1991, the Ford Motor Company sold Ford

ABOVE:
Ford special-build No. 0708, a 4000 fitted with a torque converter transmission for towing aircraft at Stansted in 1970.

CENTRE LEFT:
The Versa 6800 high clearance tractor, based on a gasoline powered version of the Ford 6600 tractor and built by F W Larson of Texas, USA.

LEFT:
A Ford 7610 fitted with a Bomford front-blade, HIAB 650 crane and an Igland double-drum winch. The conversion was built by A T Osborne Limited for forestry work.

ABOVE: Designed for use on peat, a Ford 6410 Generation III tractor fitted with Difco swamp tracks built by Ballintubber Enterprises of Ireland.

New Holland to Fiat, Italy's largest private company. Today, with the combining of the New Holland Ford and Fiatagri ranges, sales of the power-trains and components is still strong. This side of the business is now handled by the Department of Industrial Sales and Marketing at Basildon, which also deals with highway tractors and tractor loader-backhoes.

Even Caterpillar is a customer for Ford components - the new Challenger 35 and 45 machines, introduced in 1994 and built in

RIGHT: Another conversion by A T Osborne, a Ford 7810, fitted with an FMV 470 crane, supplied to Fountain Forestry in 1991.

LEFT:
The Bruun 7620F timber forwarder, built from 1990 by Firma Lars Bruun of Philipstadt, Sweden, and based on the Ford 7610. It has eight-wheel drive and a two-speed hydrostatic transmission. Designed as a self-loading transporter, it is capable of hauling 10 tonnes of sawn timber, loaded by the Fiskars hydraulic crane.

Ford's Versatile Factory in Winnipeg, Canada, source some of their transmission parts, cabs, controls and hydraulic linkage from the Ford P396 skid unit, used in the Ford Genesis 70 and Fiat 'G' Series tractors.

In January 1995, the company ceased to use the Ford oval, and a new corporate identity was established using the New Holland name with a new logo. As we reminisce over the Ford tractor conversions of the past, we can look forward to a new generation of New Holland tractor conversions for the future.

LEFT:
Several components in the new smaller Caterpillar Challenger tractors are sourced from the Ford P396 skid unit. The impressive Challenger 35 is seen on its first European demonstration in December 1994.

CHAPTER 2

County

A 1966 County 954 on flotation tyres hauling power lines for an electric pylon scheme.

RIGHT: The Market Transport Company Ford AA truck with the third-axle conversion built in the late 1920s.

BELOW: The County driven third-axle 6 x 4 conversion for the Ford AA truck, built circa 1930/31.

ABOVE: 1929 County Commercial Cars Sussex conversion of the Ford AA.

County's involvement with Ford and Fordson tractors is probably greater than any other company's. All its vehicles were built almost exclusively from Ford skid units, and it was also the most prolific manufacturer of Ford conversions. In County's first twenty-five years of tractor production, from 1948 to 1973, the company built and sold around 30,000 crawler and four-wheel drive tractors. County's connection with Ford products,

LEFT:
A 1929 Ford AA truck fitted with the County Surrey 6x2 conversion.

ABOVE:
The Sussex/Surrey conversion on a 1932/34 Ford BB truck.

however, goes back to long before the start of tractor production in the late 1940s.

The company was founded by two brothers, Ernest and Percy Tapp. They both served as army officers in the First World War, Percy involved in administration and Ernest in engineering and motor transport. After the war, Ernest, married to a butcher's daughter, was horrified to discover meat was still transported to and from Smithfield Market in primitive horse-drawn wagons. Drawing on his wartime experience with motor vehicles, he set up the Market Transport Company with his brother, operating from Weir Road in Balham, south-west London, to deliver meat in covered Ford AA trucks.

The company soon found the small 30 cwt trucks were not big enough for their needs, so Ernest set about converting one to carry two tons by adding a third trailing axle to the Ford chassis. Another version was built, this time with power transmitted to the third axle via a worm drive. The Tapp brothers were soon receiving enquiries from other

TOP: An interesting Baico extended chassis version of the 1930/31 Ford AA Sussex.

RIGHT: The County designed Ford Tug, based on the Y model car, was built in small numbers from 1935.

BELOW: Wartime County Sussex 6 x 4 conversion on the Ford WOT 1 military vehicle chassis.

transport firms wanting six-wheel trucks, so in 1929, they set up County Commercial Cars Limited at Fleet in Hampshire as a separate concern to build third axle conversions on Ford AA chassis. The County name came from the brothers' practice of naming their vehicles after counties - the trailing axle (6 x 2) conversion was called the Surrey, while the driven axle (6 x 4) truck carried the Sussex name. Ernest looked after the engineering side of the business, and Percy helped out in administration and sales, while still running the transport company.

In 1931, County improved the six-wheeler by using a double crown-wheel drive system. The Ford Motor Company adopted the design and put it into production at Dagenham on the Model A truck, County supplying the necessary conversion parts, rear axle and suspension. Production of the Sussex and Surrey conversions continued at Dagenham with six-wheel versions of the Model B and 7V trucks.

In 1935, Ford built-up a few County-designed small three-wheel tugs. A van version, known as the Devon Distributor, was

LEFT:
RAF mobile generator mounted on a Ford WOT 1 Sussex conversion.

BELOW:
Ford WOT 1 Sussex converted for medical use by the RAF.

manufactured at Fleet along with the Devon fire-engine, but not in large quantities.

During the war years, production was stepped up dramatically, and a total of 14,000 military versions of the double-drive Sussex were supplied. The majority of the 6 x 4 military vehicles were based on the Ford WOT 1 chassis, and included ambulances, fire-tenders and mobile workshops, along with more general-purpose trucks. A number of Sussex conversions were built up on the Fordson E817T chassis, which was then fitted with winches powered by the Ford V8 engine, and used for launching barrage balloons.

Even though County was

LEFT:
Sussex conversion of the Fordson Model 79 chassis used as a winch truck for launching barrage balloons. The winch is driven by a Ford V8 power unit.

operating to full production capacity by 1943, it still found time to develop a two-man tank based on a Bren-gun carrier chassis. Known as the Praying Mantis, it was designed to give close armoured support to infantry fighting in villages. An armoured compartment, fitted with two Bren-guns sighted by periscope, was elevated, allowing the gunner to fire at the enemy over walls or hedges. In the event, the machine was regarded as too specialised to be of any great use, and did not enter service.

ABOVE: The Fordson E817T Sussex barrage balloon winch truck in action.

RIGHT: The 'Praying Mantis' experimental two-man tank.

The Praying Mantis was designed at the request of the General Staff after inspecting a smaller machine operated by one man. The figures on the plating indicate the thickness of the armour. The armament consists of two Bren guns inverted and so mounted that they can be removed for use apart from the machine in a matter of seconds. They are fired by a very simple remote control; periscope and guns move in unison so that the sights are always in the gunner's view. Guns can be fired singly or simultaneously and recoil effect on the gun box is completely balanced.

The machine can be driven with the elevating compartment in any position and lowered from full elevation in 3½ seconds.

Vision to the rear is afforded by angle door reflectors at the helmet trunnion centres.

Praying Mantis in prone position with guns directed forward.

Compartment slightly elevated to normal travelling position. In this position the ride is extremely comfortable by reason of the double springing on suspension and box elevating mechanism enabling a higher speed to be maintained across country than with any cross country vehicle of comparative size.

Shows the machine in the prone position in comparison with a standing man. The picture also shows gunners' and drivers' visors which can be opened for direct vision. Driver's periscope is shown directed right.

Full elevation. In this position, as in all elevated positions, an all round traverse is possible, the gunner rolling over to fire to the rear. The gunner's periscope is below the gun box and directed with the sighting means in unison with the guns. Full width helmet can be tilted 30° up or down. Guns finally aligned by direct control.

Taking up of camouflaged position, the height being adjusted so that only the gun box and periscope are above cover.

DESIGNED AND BUILT BY
COUNTY COMMERCIAL CARS LIMITED
1943/1944

BELOW: RAF recruiting office mounted on a Ford WOT 1 Sussex chassis.

With the introduction of the new Ford Thames Trader lorries in 1957, production of the Sussex and Surrey six-wheelers ceased at Dagenham, but County, in conjunction with the Ford Motor Company, built six-wheel conversions of the Trader at Fleet. The new range of lorries was known as the County Thames Trader Tandems. Based on six-cylinder versions of the Thames Trader, the County Tandems were designed to operate at a gross laden weight of upto 16 tons, and were

LEFT:
County Thames Trader Tandem 6 x 4 conversion.

LEFT:
County Thames Trader Tandem with a body designed for the bulk handling of animal feeds

LEFT:
Ford Thames Trader 6 x 2 with the County trailing third-axle.

RIGHT:
The County double-drive rear axle conversion is clearly shown in this picture of a 1961 6 x 4 Thames Trader concrete mixer.

BELOW:
A look inside the 1949 County Full Track crawler.

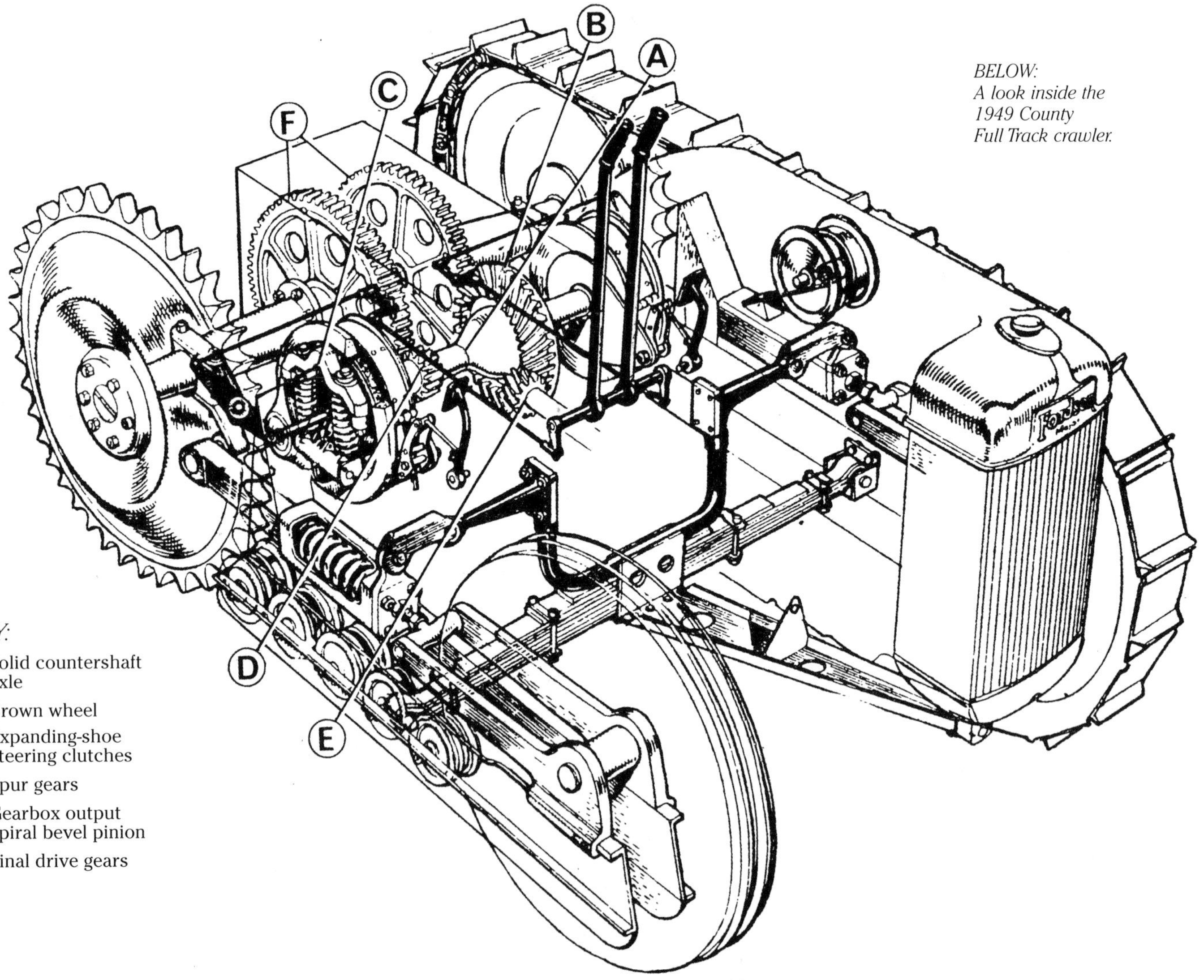

KEY:

A. Solid countershaft axle

B. Crown wheel

C. Expanding-shoe steering clutches

D. Spur gears

E. Gearbox output spiral bevel pinion

F. Final drive gears

LEFT:
The prototype County CFT at work. Note the early solid-disk front idler.

fitted with a variety of special-purpose bodies. Tipper versions were supplied with either Ford's own, or a David Brown, five-speed gearbox.

By the 1960s, production at Fleet was divided almost equally between tractors and commercial vehicles, and the company's service engineers were expected to be responsible for both products. County did continue to build some six-wheel versions of the Ford D series lorries up to the end of the 1970s, but truck production dwindled as tractor manufacture took over as the mainstay of the company's business.

County's establishment as a tractor manufacturer was an unexpected, if fortuitous, move. After the massive wartime production run for the Sussex

ABOVE:
The prototype County Full Track on trial with the Ford Motor Company in 1948. On the right of the photograph is County's chief designer, Joe Davey.

LEFT:
A 1949 County CFT fitted with a Bray angledozer .

ABOVE:
1963 County Thames Trader 6 x 4 Tandem tipper lorry.

RIGHT:
County Full Track crawler. This is an early TVO model, with pressed disc font idler.

LEFT:
Later model County CFT with the spoked cast front idler. This example is fitted with a Perkins P6 six-cylinder diesel engine and dates from around 1950.

LEFT:
The first County crawler sold in South Lincolnshire, supplied new to farmer, Ken Hatter, for the pre-decimal price of £815 13s 0d.

LEFT:
Some of the County management, and other interested parties, gather for a demonstration of the model Z crawler working with a Ransomes Dragoon disc plough.

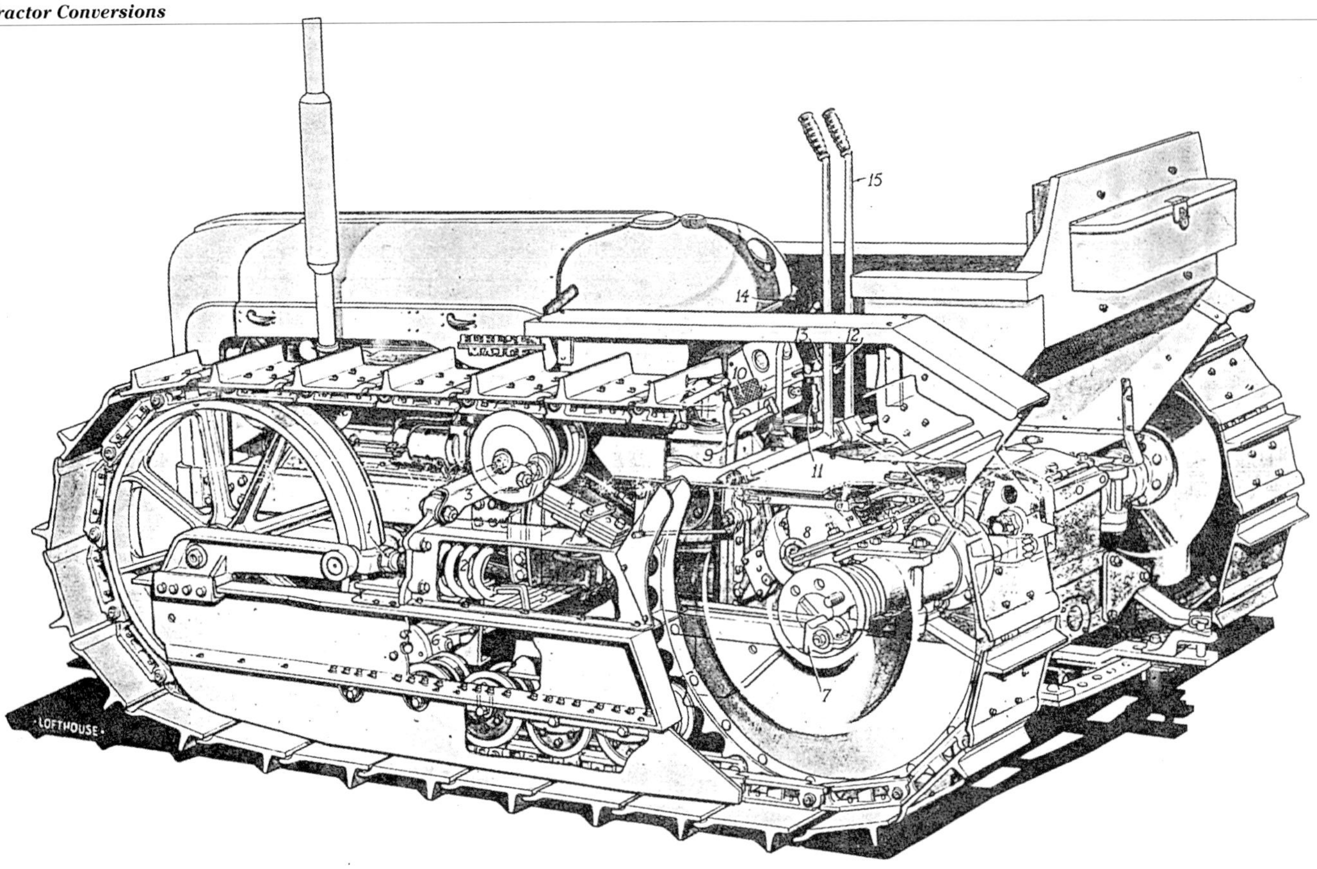

KEY TO NUMBERS ON DRAWING

(1) Track adjuster	(6) Spring anchor box	(11) Parking brake lever
(2) Recoil spring	(7) Rear support of track frame	(12) High-low speed change lever
(3) Radius rod	(8) L.H. steering clutch and brake	(13) R.H. brake pedal
(4) Top roller support spring	(9) Engine clutch pedal	(14) Gear lever
(5) Transverse spring	(10) L.H. steering brake pedal	(15) Steering levers

ABOVE: An exploded view of the County Model Z crawler. (Farm Mechanization *copyright)*

military vehicles came to an end, County were looking for new products to fill unused production capacity at Fleet. The company were building a number of spraying machines for Pest Control Limited of Cambridge. These were based on Ford V8 trucks mounted on 'stilts', with drive to the rear wheels through extended chain cases. Both 4 ft and 7 ft high clearance versions were built, designed to keep the operators well away from the chemical spray. The machines were de-signed for spraying for fly on fruit bushes with nicotine, which was deadly in its purest form.

RIGHT: County crawler fitted with a Leeford hydraulic Muledozer.

Experience with these machines led to County being asked by Pest Control to develop a narrow gauge crawler for orchard spraying. Also, the Ministry of Agriculture were looking to buy a British-made crawler tractor for distribution by county agricul-

LEFT:
County P50 Ploughman ploughing near Newark, Nottinghamshire.

enter the tractor market, albeit for what was only expected to be a short production run, and build a machine to fulfil both needs. In the event, the demand for the resulting machine was far greater than ever imagined, and heralded the start of County's long line of tractor production.

The crawler was based around the Fordson E27N Major tractor, and the prototype was undergoing trials with the Ford Motor Company and the National Institute of Agricultural Engineering by the summer of 1948. Design for the machine was simple. The Fordson tractor's differential was replaced by a solid countershaft axle, the crown wheel of which was still driven by the normal gearbox output

tural executive committees. As Ford were not interested in directly involving themselves in the manufacture of such a machine, it appeared an ideal opportunity for County to

pinion. Internal expanding Girling brake shoes, enclosed in brake drums, formed the steering clutches, transmitting the drive from the countershaft to the final drive, via spur

LEFT:
County Long Swamp tractor, based on the Mark III industrial crawler, cutting drainage channels in Scotland with a Cuthbertson plough .

ABOVE: A typical County scene from the 1950s, taken in the old works yard at Fleet. Two Ploughman crawlers are loaded for delivery on to a County 6 x 2 Thames Trader lorry.

RIGHT: County CD50 fitted with a Braydozer attachment.

gears. The steering clutch mechanism worked rather like a car brake in reverse: pulling on either hand lever would release the internal shoes from the drum, disengaging drive to the appropriate track. Foot pedals operated external contracting brake bands which further aided turning.

The new crawler had five track-rollers, with renewable steel tyres. International TD6 tracks were used on the prototype, but

LEFT:
County Ploughman, with extended stack-pipes, outside the development works at Fleet.

LEFT:
County Ploughman P55 crawler.

County developed its own pin and bush tracks for the production tractor. A large sprocket and front idler were fitted to accommodate the Ford skid unit, and give the machine adequate ground clearance. An advantage of the large sprocket was that each of its teeth only engaged with a track bush every third revolution, reducing wear and maintenance costs, and the high track profile provided exceptional traction for a small crawler.

Fifty machines were delivered to the Ministry of Agriculture by the end of the year, and the crawler was in full production for release on general sale by January 1949. It became known as the County Full Track, or CFT, and was initially priced at £765 ex works. Later that year, the CFT also became available equipped with a Bray hydraulic angledozer at a cost of £1,140 complete. Bray and County continued to have a long association, and most County crawlers and early four-wheel drives could be bought with angledozers supplied by the Feltham company.

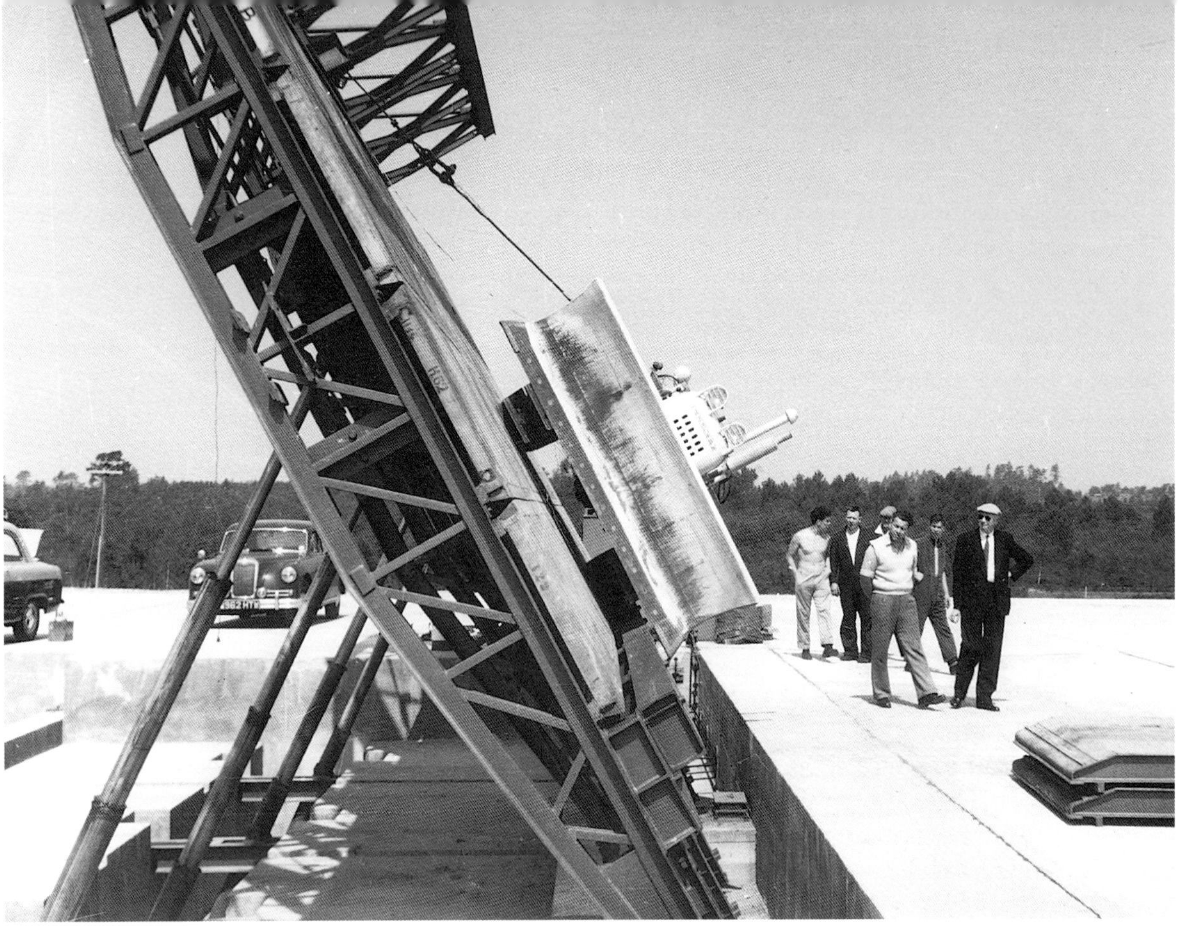

RIGHT: County CD50, fitted with a Braydozer, undergoing a turn-over test, conducted with the engine running, at the Military Engineering Experimental Establishment (MEXE) at Ringwood in Hampshire

RIGHT: County CD50 extracting timber with a Boughton winch.

OPPOSITE BOTTOM: County CD50 Refuse tractor

The 29 bhp Fordson petrol/TVO engine was found to be, at times, somewhat lacking in pulling power. It also suffered some problems with oil starvation to the front main bearing if the driver operated the machine on gradients in excess of one in four, ignoring warnings to the contrary in the instruction book. The ideal solution was the Perkins P6 six-cylinder diesel conversion for the E27N Major, introduced in 1948, and almost immediately adopted by County. The Perkins engine did not have the inherent lubrication problems of the TVO engine, and the crawler fitted with this engine found its available power boosted to 45 bhp, but the price tag of the basic machine was now £1,090.

A few improvements were made to the CFT during its production run. In August 1950, cast spoked front idlers were fitted in place of the original pressed discs, and in October 1951, the Model B CFT was released, with strengthened track frames to stand the rigours of industrial use. But the CFT's days were numbered with the E27N Major due to be replaced by a new model from Ford.

When Ford's E1A Major was released to the trade in a blaze of glory at the Kursaal in November, Percy Tapp was in attendance to answer queries regarding County's crawler version of the new tractor, which was already in prototype form. The new model Z County crawler was tested in field trials before the end of the year, and was in production by July of the following year. Production of the old CFT model did not end completely, and several were made up to 1955, mainly for export to third-world countries.

There was little difference between the County Z crawler and the CFT it replaced, apart from design changes to accommodate

BELOW:
A 1959 County CD50 with a Boughton side-boom crane, laying pipes in Hertfordshire.

RIGHT:
County Ploughman P55.

the new Major skid unit which now had the advantage of a six-speed gearbox and Ford's excellent new four-cylinder diesel engine. Petrol and TVO options were offered, but rarely fitted in the County, and when the Z evolved into the Mark II crawler in 1954, the company only produced the 40 hp diesel version.

The agricultural Mark II and the export and industrial Mark III versions were both released at the same time. The Mark III industrial machine was available as a base machine, a bulldozer with a Bray attachment, or as a side-boom crane tractor for pipe-laying, fitted with equipment supplied by Boughton.

A further interesting variation on the Mark III was the Swamp tractor with wide-gauge tracks for working on marshland, peat bogs and swamps. This machine was usually supplied in extended form as the Long Swamp tractor, with two top rollers and nine bottom rollers. The machine was extended by mounting an extra Fordson transmission housing between the gearbox and the standard transmission, with the gearbox output shaft extended through the dummy transmission. The Long Swamp tractor exerted a ground pressure as low as 2 psi on peat, and was used extensively by the Forestry Commission in Scotland for draining moorland and reclamation work.

Problems with stress fractures were encountered with the Mark III when it was used with heavy

BELOW:
The first County six-cylinder crawler, sold by W.J. Cooper, working in Bedfordshire.

equipment such as the bulldozer, so County was faced with having to brace the rear axle to the side frames. This was overcome by the introduction of the Mark IV industrial in 1956, which had heavy-duty rear axles in cast steel casings, and suffered none of the drawbacks of the previous machine.

The first major design changes to the County crawler came with the new Ploughman tractor, introduced at the end of 1957 to replace the agricultural Mark II model. The new machine had roller-bush tracks, and the option of multi-plate steering clutches in place of the old expanding shoe system. Track-life was increased, steering was easier and less maintenance was required, as the new steering clutches were self-adjusting, although County always maintained that the old expanding shoe steering system worked perfectly if kept in correct adjustment. The

ABOVE:
Extended County Long Swamp tractor, based on the CD50, fitted with a Shawnee-Poole Warrior backhoe.

LEFT:
The County Four-Drive undergoing trials in 1955.

ABOVE: This drawing clearly shows the chain-drive arrangement on the County Four-Drive. (Farm Mechanization *copyright)*

external appearance of the new crawler was altered by the provision of a heavy duty radiator guard.

The Ploughman was also the first County to have the benefit of over 50 hp, when it was uprated to the specifications of the new Fordson Power Major tractor during 1958. This model, known as the Ploughman P50, was rated at 51.8 bhp and had multi-plate steering clutches as standard.

The CD50, an industrial version of the improved crawler, was introduced in 1959, and it was available with a complete range of ancillary equipment, including the Bray angledozer, Boughton side-boom crane and winches, and scrapers supplied by Jack Olding and Conder Engineering. County themselves built a heavy duty toolbar for use with the tractor, and JCB offered their Hydra-Digga backhoe as an attachment for the machine.

County also built a special refuse tractor, with an extra large dozer-blade and modified track plates, for use on council tips with local corporations. To keep the engine dust-free, side panels were fitted, and the 'extractor' cooling fan was replaced with a blade that blew, rather than sucked, air through the radiator. The air-breather was

RIGHT: County Four-Drive skid-steer tractor hauling timber.

FACING PAGE TOP: County Four-Drive working an Onions scraper supplied by Jack Olding.

FACING PAGE BOTTOM: County Hi-Drive high clearance conversion of the Fordson Major. A County crawler can be seen on the assembly line behind.

COUNTY

ABOVE:
County Four-Drive fitted with a Braydozer attachment.

RIGHT:
Driver's-eye view of the County Four-Drive. Controls, left to right : hand clutch; foot clutch; L/H steering brake; gear lever; L/H steering clutch; R/H steering clutch; hi-low gear lever; throttle; R/H steering brake; hydraulic lift control lever for dozer blade.

ABOVE:
Rear view of the Four-Drive. The oil reservoir for the Bray equipment can be seen under the driver's seat.

LEFT:
The casing for the train of three gears, transmitting drive down to the rear axle stubs, can be clearly seen in this photograph of the County Hi-Drive. The drawbar is adjustable for height.

RIGHT: The County Hi-Drive had 30 inch clearance under the front axle.

OPPOSITE TOP: The County Super-4. The front axle drive casings are pivotally mounted on the front axle underbeam.

OPPOSITE BOTTOM: A 1961 County Super-4 test tractor, which was on permanent loan to the NIAE at Silsoe.

BELOW: A Fordson Power Major fitted with County Viking track equipment working on the beach.

fitted with a special pre-cleaner with a dust collector, and the operator was protected by a Winsam cab.

County uprated the Ploughman agricultural crawler still further in the early 1960s, by using the Super Major skid unit fitted with a 55 hp industrial version of the Ford engine. Designated the P55, it had heavier tracks and a larger front idler to improve the angle of the tracks as they came into contact with the ground. It was finished in the same lemon yellow colour scheme as the CD50 and priced at £1,855.

By this time, the company's quality control standards were high, and it had its own engine test and development department. Each engine received from Ford was dynamometer tested and the injection pump was recalibrated to meet County's requirements. County also built and fitted its own heavy-duty crownwheel, pinion, and final reduction gears.

Every County tractor sold was

SUPER MAJOR
SUPER-4

840
DCC

visited and inspected by a member of County's team of service engineers within six months of purchase, when any necessary adjustments were carried out. However, the engineer covering the eastern counties ran into some difficulties when visiting the Geest farms near Spalding in Lincolnshire, to inspect three P55 Ploughman tractors sold to the company a few months previously. Concerned at being able to find only one crawler to inspect, he made some enquiries about the whereabouts of the other two

RIGHT: The County Super-4 on test in the1960s.

OPPOSITE TOP: The solution to difficult conditions - but not recommended by the company. A County Super-4 on steel wheels could lead to some expensive component failures.

RIGHT: Four driven wheels will spin! A County Super-4 finds the going heavy with a six-furrow plough.

OPPOSITE BOTTOM: A Super-4 clearing snow from County's old yard in the winter of 1962. The company's Ford service van has been sensibly left in the shed.

machines. He was asked if he could swim, and he said excitedly that he could, expecting to be invited to the Van Geest family's new swimming pool, but his hopes were dashed when he was told that he would have to swim an awful long way, because the missing tractors had been sent out on the company's banana boat to work in Trinidad !

The P55 Ploughman was not quite the last version of the County crawler. A basic no-frills model, known as the Clydesdale, was produced for a while, fitted with the standard 52 hp Super Major engine. Originally envisaged as a cheaper export model, there was little demand for it and very few were sold.

County also brought out a six-cylinder Ploughman in 1963, fitted with the Ford 590E engine. Again very few were built, but Ford dealers, Gates of Baldock, continued to convert a few of their customers' own

crawlers over to six-cylinder engines.

County crawler production came to an end in 1965. Demand for the crawler still existed, but not as strongly as before with new four-wheel drive tractors on the market. Ford's new 5000 tractor was now coming off the Basildon assembly lines, and County were unwilling to invest a lot of money in devel-

RIGHT: County Super-4 fitted with a Sapper Powerdozer.

BELOW: County Super-4 logging in Africa.

oping a crawler to fit this totally new skid unit. County approached the Italian crawler manufacturers, Mailam, to build a suitable transmission housing, but the cost involved was uneconomical. The company was not overjoyed when Gates, the second largest County dealer, started importing Mailam crawlers, based on the Ford 5000, a few years later.

It was said that the last County crawler built was a Long Swamp model which went to the Forestry Commission in Scotland, where it worked until being sold in 1994. After crawler production ceased at Fleet, the company concentrated on its expanding four-wheel drive tractor market, but to quote from County News, 'It was not without some regret when we at County ceased to be Crawler men.'

County's first foray into the four-wheel drive market was almost accidental. By the 1950s, the company was slowly moving into the export market, and had a steady trade supplying crawlers to the sugar cane industry in the West Indies. In Puerto Rico the crawlers were used on large estates for field extraction, hauling the cane to railway systems which took the crop to the refinery. Eventually, central refineries were built and the railway systems were scrapped in favour of road haulage, and the local County agents were asked to supply a tractor that had the traction of a crawler, yet could be driven on hard roads. To the County company back at Fleet, the simple solution seemed to be to take

LEFT:
The County Super-6 six-cylinder tractor.

BELOW:
County Super-6 outside the company's head office in Fleet.

ABOVE:
The distinctively styled County Super-6 six-cylinder tractor.

RIGHT:
Top sectional view of the County Super-6 showing the drive arrangement from the bull-pinion shafts to the front wheels.

LEFT:
County Super-6 fitted with a ground-anchor for drainage work.

BELOW:
The County Sea Horse undergoing water trials.

RIGHT: County Super-6 working with a Bomford Superflow cultivator in Bedfordshire.

the crawler and put it on rubber tyres, which is exactly what they did.

Introduced in 1954, the County Four-Drive was described as a wheeled crawler. The tractor was basically the crawler skid unit fitted with four equal-size wheels. The front wheels were chain driven from sprockets mounted on the rear-wheel axle. The chains ran in an oil bath inside articulated chain cases, which also housed the front-wheel hubs. The normal expanding shoe steering clutches were retained, and the tractor was 'skid-steered', relying on flex in the pneumatic tyres allowing it to turn. Thus County's first four-wheel drive tractor was born.

BELOW: County Super-6 hauling an Onions scraper in Iran.

The Four-Drive met all the requirements of the sugar cane industry, and could haul four or five laden trailers at speeds up to 14 mph with ease. Puerto Rico was to eventually become such an important export market for County four-wheel drive tractors that later the company suggested, 'If you are going to be run over by a tractor in Puerto Rico, it will almost certainly be a County!'.

Unfortunately, the Four-Drive had no real application on the home market. The company tried to encourage sales by offering it with a timber winch

LEFT:
A County Super-6 tractor fitted with a Waldon angledozer. A number of these American attachments were imported by Opico Limited.

for forestry work, but most of the few sold in the UK were fitted with Bray angledozers to the construction industry. A few were supplied in skid-unit form to Muir-Hill who based their 4WL loading shovel on it. The majority of Four-Drives, however, went overseas to the West Indies, South America and the Far East. The Four-Drive was fitted with multi-plate steering clutches from 1958, and still offered by the company up to 1962, but not many were sold after the more practical County Super-4 came out.

ABOVE:
David Tapp on the County Sea Horse after having crossed the Channel in 1963.

LEFT:
County Sea Horse undergoing trials on water.

The County Hi-Drive high clearance conversion of the Fordson Major tractor was another product initially built at the request of sugar cane growers in the West Indies. The front axle was altered to give 30 inch ground clearance, and the rear was raised by inserting a vertical train of three gears down to the wheels which were mounted on stub axles. Released to the UK in August, 1958, it was quickly taken up by celery and potato growers; it also became a popular spraying tractor. County

RIGHT: David Tapp puts the Sea Horse to work with a Cuthbertson drainage plough.

would supply it as a complete machine, or as a kit to convert existing tractors.

By this time, County's product line included imported flexible half-track equipment to fit the Fordson Dexta and Major tractors, known as the County-Viking track.

BELOW: Mr David Tapp in the drainage channel cut by the Cuthbertson plough.

The company also manufactured the Highway Dexta from 1960 to 1964, built to comply with UK road regulations for the needs of local authorities. New export markets were opening up, the crawlers and commercial vehicle conversions were still selling well and new wheeled tractors were in the pipeline.

With the need to expand, County moved across the road in Fleet from the corrugated-iron clad buildings of the old works, to a new purpose-built factory and office complex, opened in 1955. Also, a new company, County Commercial Engineers Limited, was started up supplying precision machined tractor parts to the parent company from premises in South Norwood, London, previously used by a sister firm, Radio Electronics.

Ernest and Percy Tapp's sons had also entered the business. Percy's son, Raymond, concentrated on the transport and commercial vehicle side, eventually becoming joint managing director of the company, before taking over from his father as chairman. Geoffrey, one of Ernest's two sons, working through all the production processes, became technical director with responsibility for engineering, and then finally joint managing director with Raymond. Geoffrey's brother, David, looked after technical liaison and special applications before becoming

LEFT:
County heavy duty drawbar attachment for the Super-4 tractor.

sales director.

One of the new machines in the pipeline was the Super-4, brought out in 1960, based on the Fordson Super Major. This was a totally new four-wheel drive design, with power to the front wheels transmitted by twin propeller-shafts driven by bevel gears off the Fordson bull-pinion shafts. This system allowed the normal differential to lock all four wheels, and was to be a feature of almost all designs for County tractors in the future. The conversion relied heavily on standard Ford parts. The front wheel drive casings contained Fordson Dexta crown-wheels and pinions, and the propeller shafts were built from Thames Trader components.

Competitively priced at £1,470, the Super-4 soon became popular with farmers wanting the advantages of a four-wheel drive tractor. Although less powerful than the Doe Triple D, Matbro Mastiff and Bray Centaur four-wheel drive tractors, it was nearly £1,000 cheaper, and infinitely more compact.

To compete with these higher horse power machines, County introduced a six-cylinder version of their tractor, designated the Super-6, and powered by the 95 bhp Ford 590E engine. The agricultural model had a fibreglass bonnet and a distinctive radiator shroud, whereas the industrial version used normal Ford tin-work, as the new styling would foul front-mounted attachments such as the angledozer.

To quote from County's advertisement for the tractor, 'The Super-6 is the same as the Super-4 only more so.' And so was the price:

LEFT:
A County Super-4 skid under construction in the factory, intended for use in an industrial machine as no hydraulics are fitted. The underbeam front axle and pivot points for the front hub casings can clearly be seen.

RIGHT:
County Super-4 skid unit. County made up its own industrial wheels.

BELOW:
Schaeff loading shovel based on the County Super-4 skid.

ABOVE: Christiansen loader based on a reverse-drive County Super-4 skid unit.

in 1964, the Super-4 cost £1,572, while the Super-6 was priced at £2,130.

The County Sea Horse was a further interesting variation on the Super-4. Designed to be capable of working in water, swamps, bogs and other similar conditions, it was fitted with flotation tanks front and rear. Extra buoyancy was provided by watertight compartments in each wheel, sealed with boiler plates. The engine and transmission were waterproofed, and all grease nipples and breather vents were extended clear of the waterline. No electrics were fitted, and the engine was started by an inertia starter. It was propelled in water by the normal motion of the oversize Goodyear tyres.

On 30 July 1963, in an unrivalled publicity stunt, David Tapp crossed the English Channel with the tractor to earn himself a place in the *Guinness Book of Records.* The crossing of 28 nautical miles, from Cap Griz Nez in France to St. Margaret's Bay in Kent, was accomplished in 7 hours 50 minutes. Within minutes of landing, the tractor further illustrated its versatility by ploughing a field near Deal.

Several Sea Horse tractors were sold for land reclamation work in East Germany, and a few were used for seismographic work for off-shore oil and gas exploration in the UK. It was normally supplied without the front and rear floatation tanks, and cost £2,266. The tanks, when required, added an extra £94 to the price.

County also sold the Super-4 as a basic four-wheel drive skid unit to other specialised manufacturers. The Danish company of C. H. Christiansen and German-based Schaeff both built loaders based on reverse-drive Super-4 skid units. The Schaeff machine was marketed in Europe under the Ford name. County even started to develop its own loading shovel, but it never went beyond prototype stage and was dropped pending production of the Super Major coming to an end.

Another interesting development of County's that never progressed beyond the prototype stage, was a pivot-steer timber tractor based on the Dexta. The machine was evolved along basic guidelines set by the Forestry Commission, who had issued their requirements for such a machine to most major manufacturers. In the event, Roadless's concept for the machine was preferred by the Commission, and they won the eventual contract with the Logmaster. County shelved its machine, and issued an article stating that there was no future for pivot-steer in agriculture.

In November 1964, the first County four-wheel drive tractor embracing Ford's new Basildon-built 6X range was released. Still keeping the Super-4 name, it was based on the 65 bhp 5000 tractor, and available with eight-speed manual, or ten-speed Select-O-Speed transmissions. Several design changes were made to incorporate the new skid unit, and drive for the front wheels, via the twin propeller shafts, came direct from the rear axle through bevel gears mounted on the differential output shafts.

For a time, County marketed the tractor as the Super-4 5000, but this name was

RIGHT: Prototype County Super-4, based on the Ford 5000, built in 1964.

RIGHT: County Super-4 654.

LEFT:
County 654 fitted with a Bomford Sapper Powerdozer.

LEFT:
County 654 fitted with a Bomford loader.

RIGHT: County 754 fitted with an NIAE self-steering device for on-the-land ploughing.

RIGHT: Export model 754, designated County 5000-Four, probably destined for the USA.

ABOVE:
County 754 working a Mather and Platt pea viner.

LEFT:
The County Super-6 954 pictured outside the offices at Fleet.

RIGHT: County 754 fitted with a Bomford sapper loader.

eventually dropped in favour of 654: 65 denoted the horsepower and 4 was for four-wheel drive. This heralded the start of County's practice of designating each model numerically. The last number indicated whether the machine was two- or four-wheel drive, and the first figures closely represented the horsepower. Hence the new Hi-drive conversion of the 5000 was

RIGHT: Top view of the County 954, showing the bevel drive for the prop-shafts taken off the differential output-shafts in the rear axle.

LEFT:
County 954, fitted with a Sapper Powerdozer built by Bomford and Evershed Ltd.

BELOW:
An early 1965 County 954 prepares for work in Yorkshire with 15 ft Cooper discs and a Massey Ferguson seed drill in tandem. In the background is a 1961 Ford Thames Trader.

ABOVE:
A 1968 County 1004 with a Wickham Poole semi-trailer used for pipeline construction.

RIGHT:
County 1124 on trial with Dowdeswell Engineering.

LEFT:
Another view of the 1124 showing the Dowdeswell DP1 plough.

known as the 652H. With the introduction of the Ford Force range in 1968, the 5000 was uprated to 75 bhp, and the four-cylinder County tractors became the 754 and 752H respectively.

The Super-6 based on the Super Major continued in production well into 1965. The Super-6 954, the first County six-cylinder tractor to be based on the 5000 skid unit, was introduced at the Royal Show in July 1965, and fitted with the Ford six-cylinder 2703E 95 bhp engine. The company was justly proud when the 954 was awarded a gold medal from the USSR All Union Chamber of Commerce for design and performance in 1966. The 954 was only manufactured for two years before being modified and re-designated as the 1004, with a strengthened rear axle, and its engine re-calibrated to give 100 hp. A further six-cylinder County

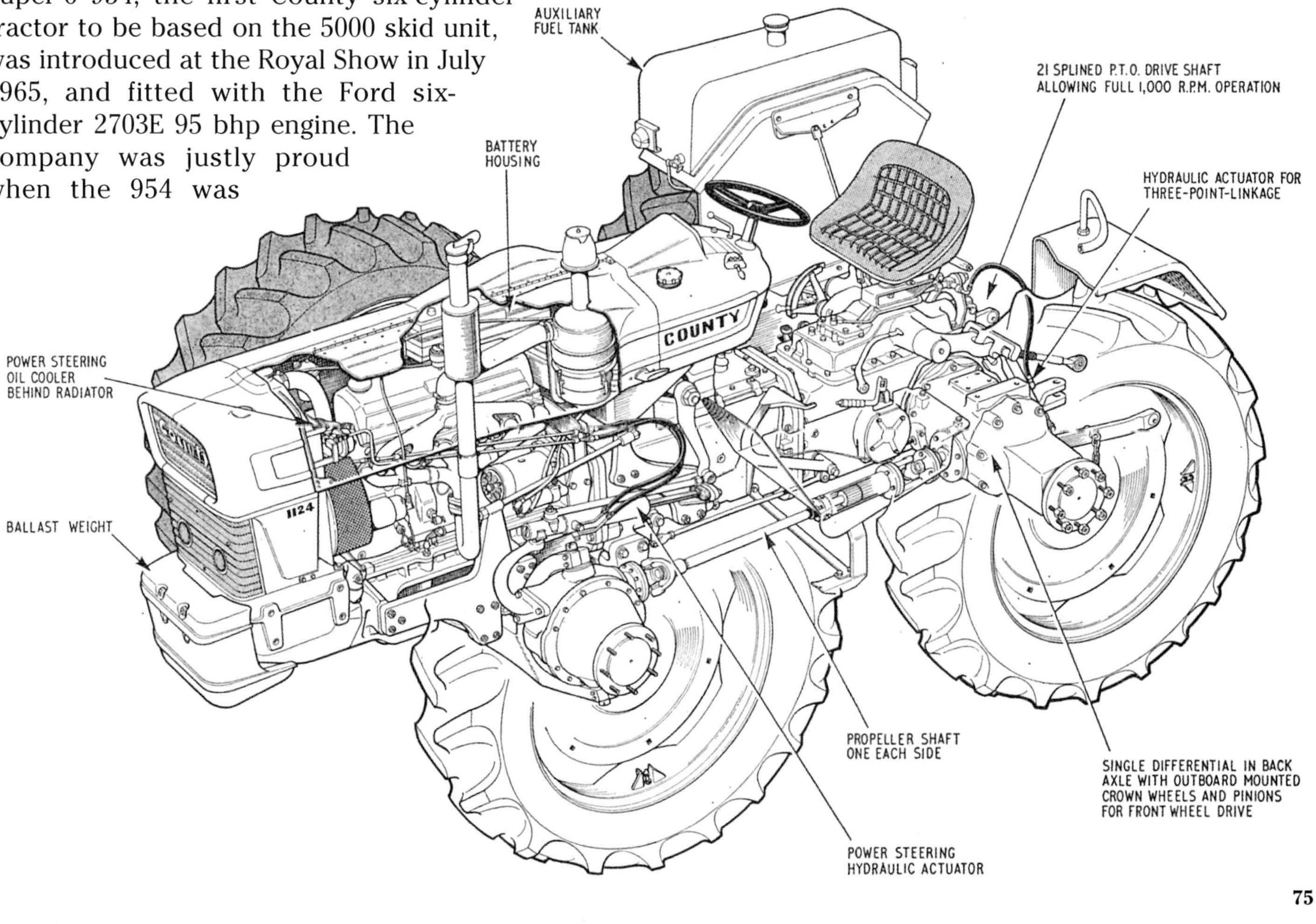

BELOW:
Exploded drawing of the County 1124. (Power Farming *copyright*)

ABOVE:
The back furrow has been removed from this five-furrow Lemken plough to allow the County 1124 to cope with the heavy clay. This tractor has been fitted with home-made auxiliary fuel tanks.

RIGHT:
County 1124 on dual wheels hauling two-and-a-half ton water pipes on a Wickham Poole trailer in Northern Ireland.

made its debut at the 1967 Royal Show, the 113 bhp 1124 tractor, powered by the 2704E engine, fitted with a reinforced cast sump.

Another new County made a brief appearance on the market in 1967. The 854T was the 654 fitted with a CAV turbocharger boosting engine power to 85 bhp. The tractor was designed in conjunction with CAV, and to guard against overheating problems, the tropical cooling kit, consisting of an eight-blade fan and a larger radiator with an 11 psi pressure cap, was fitted. Ford refused to guarantee the engine, so County offered its own twelve months warranty. It is not known how many were sold, but it was soon dropped.

The year 1968 saw the company's first departure from the equal-size four-wheel drive configuration, with the introduction of the County 4000-Four based on the Ford 4000 62 bhp tractor. Fitted with smaller front wheels, it was available as a conversion kit at £688, or as a complete machine at £1,780. Drive to the front axle was by a single propeller shaft taken off a transfer box

LEFT:
The Malvern Skidder, based on a County 4000-Four and built for the Forestry Commission by JHB Implements of Thetford, Norfolk.

mounted on the rear transmission housing. The propeller shaft was side-mounted to aid row-crop work and keep ground clearance to a maximum. The front axle contained a Ford 3000 differential and had an automatic over-run differential lock.

Without a doubt, the most revolutionary designed machines to come out of the Fleet works in the 1960s were the forward control machines. First off the drawing board in prototype form in early 1965 was the four-cylinder version based on the 654, which went into production as the FC 654. The spacious new cab was moved forward of the engine to allow the driver excellent visibility. The fuel tank was also repositioned behind the cab and above the engine, leaving an unobstructed platform area for mounting various bodies or equipment. Obviously aimed at the industrial user, it could also be used as a normal agricultural tractor and for spraying, fertiliser spreading or lime application.

In 1968, it was uprated to the FC 754, and

BELOW:
County FC 654 forward control tractor on flotation tyres.

RIGHT: The prototype six-cylinder County FC undergoing field trials in Northamptonshire.

a six-cylinder version, the FC 1004, was introduced. The machine was used in a variety of industrial applications and was fitted with many different attachments including cranes, drilling rigs, mobile welders, tree planters and special dumper bodies.

The National Institute of Agricultural Engineering at Silsoe developed an interesting application for the County FC as a mobile tyre test rig. This machine was fitted with two independent hydrostatic transmissions, one to drive the tractor, and one to power the test wheel mounted in a cradle behind the tractor. The County based machine replaced an earlier wheel tester, dating from 1954, powered by an International Farmall M, with the test wheel electrically driven.

BELOW: County FC 1004 fitted with a 300 gallon Chafer sprayer.

During the 1970s, production of County tractors was at a peak, and overseas orders accounted for 70 per cent of sales. When an 1124 rolled off the production line at Fleet on 31 December 1970, the company had built exactly 2,000 tractors in twelve months, and had several new models on the drawing board.

Introduced at the 1971 Royal Show, the 1164 replaced the 1124, but still used its strengthened 5000 based gearbox and rear transmission. The new tractor was fitted with a Basildon-built version of the the 6.5 litre American Ford 8000 tractor

ABOVE:
County FC 1004 ploughing with a three-furrow Ransomes reversible.

LEFt:
County FC 654 fitted with a Hiab two-ton hydraulic crane.

RIGHT: A mobile drilling rig mounted on the County FC chassis.

engine, de-rated to 116 bhp. This eliminated the need for the structural sump required by the industrial engine as fitted to the earlier model. The 1164 featured fresh styling, and was the first of a new generation of County tractors built with an integral cab. The cab was designed by Swedish cab manufacturers, Hara, and built in the UK under licence.

The 944, also introduced in 1971, was based on Ford's new turbocharged 7000 tractor. Similar in styling to the 1164, but fitted with a Duncan cab, it was rated at 94 bhp, and slotted into the range just under the 1004.

The top end of the County range was further extended in 1972 by the introduction of new 125 and 145 bhp machines for the Royal Show. The 1254 and 1454 models were based on the American Ford 8000 and 9000 tractors, but with the transmission and rear end extensively modified and strengthened

BELOW: A Vermeer tree transplanter mounted on a County FC tractor working in Richmond Park, London.

ABOVE:
The NIAE mobile tyre test rig based on the County FC tractor.

LEFT:
'On the land' ploughing with a County 1164 and a Ransomes TS106 reversible.

ABOVE: Later modified version of the NIAE tyre test rig, still based on the County tractor.

RIGHT: Introduced in 1971, the County 1164 is seen working a Bomford Superflow chisel plough.

LEFT:
A prototype County 1254 working a five-furrow Dowdeswell plough.

LEFT:
Demonstrator John Hull at the wheel of the prototype County 1454 undergoing development trials with a Dowdeswell plough.

ABOVE:
A County 1454 working in Nigeria with a County disc plough. The local dealer's Ford Transit service truck is in attendance.

RIGHT:
The County head office and works at Fleet, mid-1970s.

by County to meet the demands of a four-wheel drive machine. Both tractors had six-cylinder engines, turbocharged on the 1454, and 16-speed dual-power transmissions. They were priced at £6,310 for the 1254 and £6,770 for the 1494, and orders came in quickly, as farmers were looking for larger four-wheel drive tractors that had the ability to meet the increased pto–horsepower requirements of modern machinery.

County tractors were finding uses in many fields other than agriculture. The FC forward control tractor was already well established in the industrial and construction industry, not only as a base machine fitted with special bodies and equipment, but also as an off-road prime-mover for heavy trailers. The other tractors in the County range were also often developed into specialist machines to meet customers' specific needs. Flotation tractors

for swamp and reforestation work, timber and logging machines, aircraft tugs, loaders, snow-ploughs and blowers, even life-boat launching tractors, were all based on County units, either built by the company itself, or in conjunction with other manufacturers. The design and manufacture of some of the specialist vehicles was undertaken by a subsidiary company, County Power Drives Limited, who also marketed ancillary equipment for

LEFT: The main assembly line for County tractors in the No. 1 works building.

LEFT: The final inspection area outside No. 3 works building. The tractors on the left are prepared for export, while the three 1164s on the right are for the domestic market.

LEFT: Part of County's vehicle fleet of the 1970s. Ten Ford D series lorries line-up with a Transit service van.

ABOVE: The County 1454 tractor with a Dowdeswell DP3 plough under trial.

RIGHT: County 1454 with a Dowdeswell DP6 plough.

Ford and County tractors, including Howard reduction boxes and Brockhouse torque converters.

County also supplied a few four-wheel drive conversion kits to other tractor manufacturers. An International B-614 four-wheel drive was built up by County as an engineering exercise, and eventually International Harvester put the machine into production at its factory in Doncaster, with County supplying the necessary parts for the four-wheel drive. The production model was based on the later 634 tractor, and known as the International 634 All Wheel Drive. A County version of a Massey Ferguson was developed but never put in to production. The Leyland 485 and 4100 four-wheel drive tractors were built up from County parts supplied for fitting at British Leyland's Bathgate factory. Sanderson Forklifts also converted their machines to four-wheel drive by using the County 4000-Four front axle.

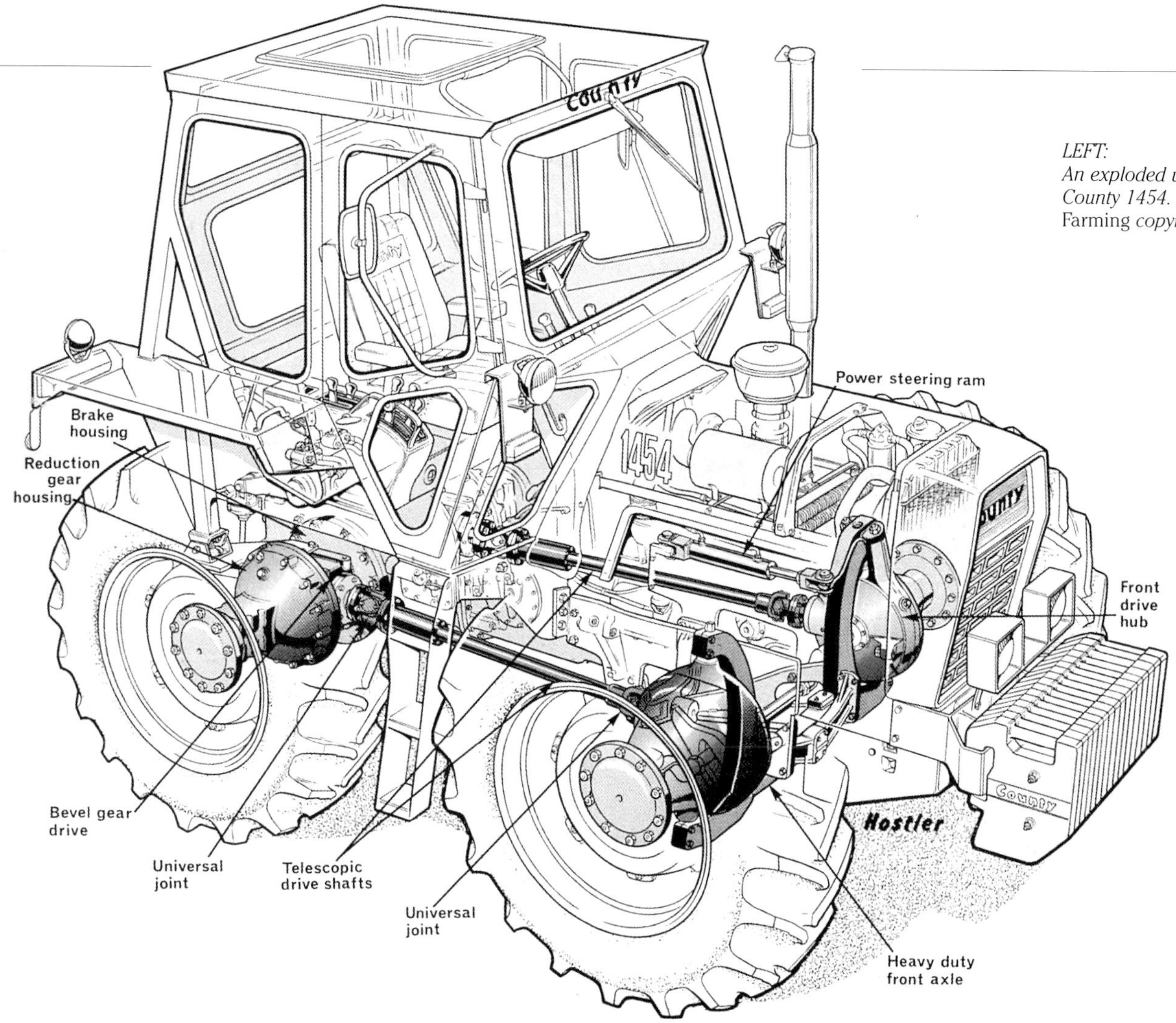

LEFT:
An exploded view of the County 1454. (Power Farming *copyright)*

LEFT:
The production model County 1454 with the Hara cab.

RIGHT:
County 1004 Flotation tractor on 42 inch wide tyres at work with the Forestry Commission.

BELOW:
County 1124 Aircraft Tug, fitted with a Brockhouse torque converter transmission, built for Hawker Siddley.

BOTTOM:
Bomford Sapper loader, built around the County 754 fitted with a Sun-Trac cab.

As demand for County tractors rose, the company found the need for further expansion, the Fleet works was extended and new offices were built. The premises at Aldershot, which housed the engineering subsidiary that had been re-located from South Norwood in 1968, also acquired more floor space. By the time County's total tractor production reached the 30,000 mark in 1973, the company was exporting to 143 different countries of the world. These were boom times for County, yet the company's fortunes were to change dramatically over the next ten years, and by 1983 it would be in receivership.

With the introduction of Ford's new 7A range in 1975, County took the opportunity

LEFT:
County 754 with a Bomford snow-plough attachment.

BELOW:
Highlander forestry tractor built by James Jones, based on a County 1164. Fitted with a Highland Bear timber crane, the tractor is still at work in the Lake District.

BOTTOM:
Canadian Pichet snow blower based on a County 754 fitted with a Select-O-Speed transmission.

to introduce new machines, and revise the existing models in the range, fitting quiet cabs in line with new UK regulations. The revamped range was introduced to the public at the December Smithfield Show.

Two new four-wheel drive tractors with unequal size wheels were now offered, based on the Ford 6600 and 7600 models. The front axle drive for the 6600-Four and the 7600-Four tractors was taken by a single propeller-shaft directly from the rear axle casing in a similar system to that used for the equal size wheel machines. The three-cylinder 4000-Four model was upgraded to the 4600-Four, utilising the new Ford model.

The other County tractors were also restyled and brought in line with Ford's new models. The 754 was replaced by the 764, and the 944 by the 964, now based on the Ford 6600 and 7600 respectively. The 1164 now incorporated the 7600 gearbox and rear transmission, and the 1454 was upgraded to 9600 specification. The 1004 and 1254 models were dropped, but the FC 1004 continued to be built. Two high clearance models were now offered, the 762H, and the 962H based on the 7600.

The company announced further changes for 1978. With the introduction of a new flat-deck cab, the 764, 964 and 1164 models were superseded by the 774, 974 and 1174 tractors. The 64 designation was retained for basic export models supplied without cabs. The

TOP: Canadian Pichet Snow Blower built on a County 754. An auxiliary engine of up to 220 hp, mounted at the rear of the machine, powered the cutting augers.

ABOVE: International B-614 with the County four-wheel drive conversion. This is the development machine, as the production model was based on the 634 tractor.

RIGHT: The drive arrangement for the County four-wheel drive system fitted to the Leyland 485 and 4100 tractors.

unequal-size wheel 6700-Four and 7700-Four machines were introduced with flat-floor cabs, while the 6600-Four and 7600-Four tractors had the Ford straddle cab.

The new forward control tractor, based on the 1174 and brought out at the end of 1977 to replace the ageing FC 1004, was easily the most interesting of County's 1978 line-up. The most novel aspect of the new machine was the modular driver's cab mounted on the right-hand side of a platform, forward of the engine, leaving an unobstructed load carrying space over the wheels. The cab was completely self-contained, with 'plug-in' electrical and hydraulic services and controls worked by Bowden cables. Fuel tanks were located between the wheels to save space and improve weight distribution. The FC 1174 was designed to carry over seven tons, with extra carrying capacity on the front platform, and could be fitted with normal hydraulic linkage at the rear, if required. Price for the machine in 1978 was £14,474.

The County 1474 was the replacement for the 1454, and was introduced at the 1978 Smithfield Show. Powered by a 149 hp turbocharged engine, it benefited from a longer wheelbase giving it increased stability. The new County was similar in styling to the Ford 9700, on which it was based.

There were mixed feelings at County as the decade came to a close. The company proudly celebrated its Golden Jubilee in 1979,

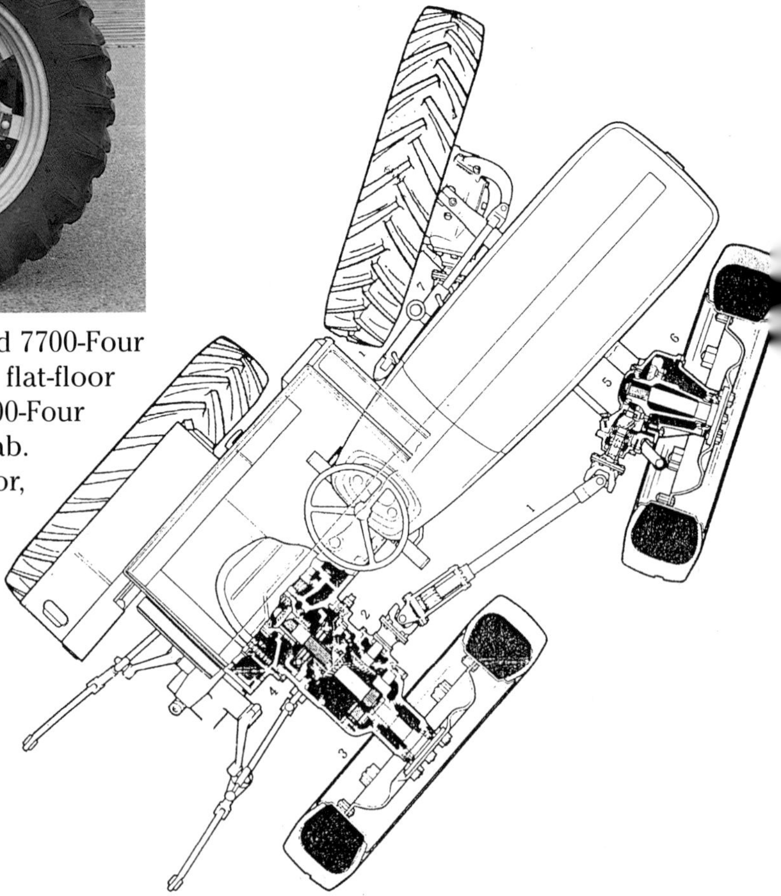

1. Telescopic drive shafts universally jointed at both ends. Twin shafts transmitting full power to to each front wheel.
2. Bevel gear drive to front wheels.
3. Epicyclic final drive.
4. Multi-plate wet disc brakes operating on front and rear wheels.
5. Spiral bevel crown wheel and pinion.
6. Individual casing pivotally mounted for steering.
7. Full power steering, hydrostatic type.

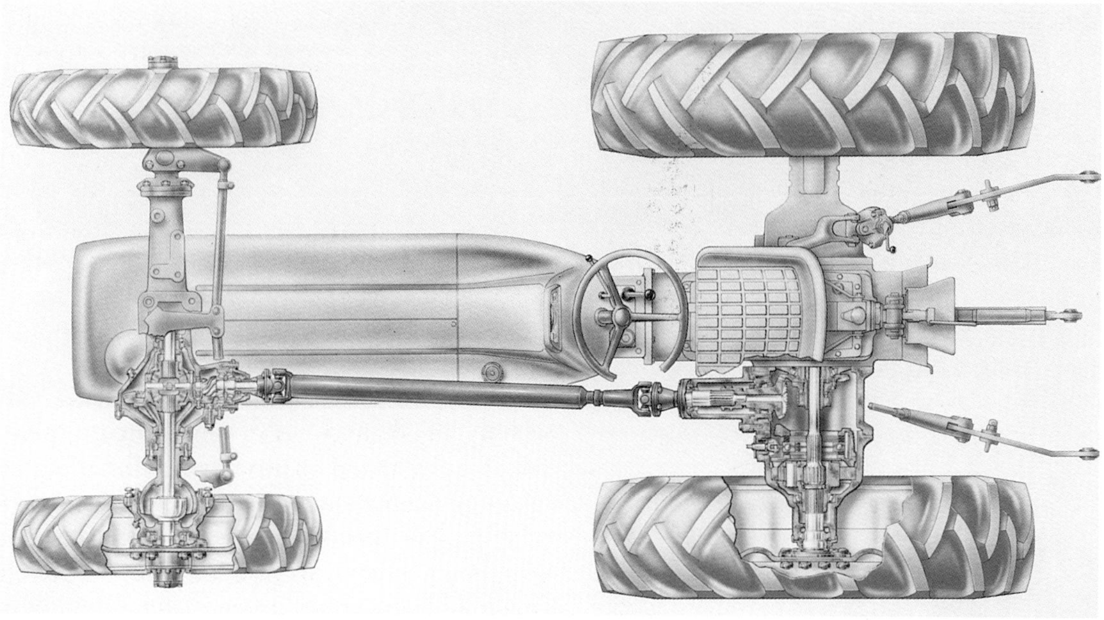

LEFT:
Top view showing the drive arrangement for the County 6600-Four and 7600-Four tractors.

BELOW LEFT:
The Sanderson rough terrain fork-lift fitted with the County four-wheel drive axle.

BELOW:
County 962H high clearance tractor on flexible half-tracks.

and an 1174 was specially finished in gold livery for promotional work. Fifty years of production at Fleet had moved from humble beginnings with early truck conversions to the manufacture of a range of sophisticated 60 to 140 hp tractors exported all over the world, with more new models in the offing. Although County was riding on the crest of a wave in the early 1970s, the storm clouds were beginning to gather.

World recession was hitting the company's important export trade, and high interest rates were affecting the home market, forcing farmers to turn to the increasing choice of cheaper four-wheel drive imports offered by foreign manufacturers.

Probably, the single most devastating blow to County was Ford's decision to introduce its own four-wheel drive tractors. Four-wheel drive versions of the four-

RIGHT: Leyland 485 with the County four-wheel drive system.

BELOW: County 4000-Four with a Dowdeswell plough

BOTTOM: County 7600-Four with a Bettinson direct drill.

cylinder 5600, 6600, 6700, 7600 and 7700 tractors, with axles supplied by ZF, had been available in Europe since 1978, and County could see it was only a matter of time before they would reach the UK. By 1980, Ford was also offering the Carraro four-wheel drive axle on its three-cylinder 3600, 4100 and 4600 tractors. The new Ford TW range of high horsepower tractors introduced in 1979, with four-wheel drive option, also hit County's sales. It was of no consolation to County that its nearest competitor, Roadless, was equally affected.

Undeterred, County faced the future with optimism, believing there was still a market for its true four-wheel drive equal size wheel machines as opposed to main-stream manufacturers' front-wheel assisted systems. In 1979, the 1174 was replaced by the 120 hp 1184TW, based on the Ford TW10. The FC 1174 was also superseded by the FC 1184TW, and the 1474 was fitted with the 153 hp turbocharged TW20 engine.

The last and largest new model equal size four-wheel drive tractor to be built by County Commercial Cars Limited was unveiled at the 1980 Smithfield Show. The County 1884 was based on the Ford TW30, and powered by its 188 hp, turbocharged and inter-cooled six-cylinder engine. The price was set at £29,850.

County continued to build new models to exploit its extensive export markets. The 1164, an export version of the 1184TW, joined the existing 764 and 1464 models, all basic machines with eight-speed gearboxes and no cabs. The 1162

LEFT:
County 7600-Four fitted with a Brown's front-end loader.

was a two-wheel drive version of the 1164. Introduced in 1980, it was designed as a simple and inexpensive tractor for field cultivations and haulage work. A high clearance 1162H was also offered for sugar cane work.

In 1982, the company introduced two tractors powered by Perkins engines for special export markets, specifically South Africa, where it was preferred to the Ford powerplant. Ford transmissions and styling were retained, but some of the machines were finished in red and green colour schemes. The 764P was powered by the four-cylinder Perkins A4.248 engine, while the 1164P had the six-cylinder A6.354 power unit. Eventually, the 1162 also became

ABOVE:
A County 1174 with a Star slurry tanker spreads sludge for the Thames Water Authority.

LEFT:
You can with a County! An 1174 demonstrates its ability to work steep slopes in New Zealand.

RIGHT: County 1174 with a Dowdeswell DP7 plough.

available with a Perkins engine.

County sales picked up slightly in 1981 as the company explored new markets and introduced fresh products, as well as continuing to build machines for special applications, including aircraft tugs, shunters and tractors for the mining industry. The company started marketing two low ground pressure vehicles, the Metrac and the Muli, built by the Austrian Reform company, and also ventured back into the commercial vehicle market with a four-wheel drive version of the Ford transit van.

RIGHT: The forward control County FC 1174 tractor introduced in 1977.

LEFT:
County 1174 about to launch a partly built craft at Vosper Hover Marine in Southampton. This tractor was finished in gold for County's fiftieth anniversary in 1979.

The County transit van was launched at the 1981 Royal Show. Fitted with an imported American front-axle and a low ratio gearbox, it was sold through Ford truck dealers. Around 120 to 150 were eventually built at Fleet, and production continues today at Knighton.

County also started development work on four-wheel drive versions of the Cortina P100 pick-up and Cargo trucks, as well as the Tractruk based on the Ford D series with increased off-road performance for slurry spreading, but these projects were eventually all shelved. The company was further involved with commercial vehicle production in 1981, assembling 31 four-wheel drive

BELOW:
This County 1174 was one of a number sold to the British Airports Authority for keeping runways clear of snow.

RIGHT: A County 7700-Four with the flat deck cab.

BELOW: A County 764 in Tanzania with a County tool-bar, which was later made under licence by Ransomes.

airport fire tenders, destined for Indonesia, for Unipower, based at Thames Ditton.

In October 1981, the County four-wheel drive tractor range was restyled to fit in with Ford's new 10 Series. Two new cheaper 'economy' versions of the unequal-size four-wheel drive range were introduced, the County 6610-Four E and the 7610-Four E, fitted with an Italian Carraro axle which had the advantage of a better turning circle than the County axle.

Orders were still coming in, and in 1982, the company supplied eleven tractors to the Ministry of Defence for shipment to Port Stanley to move supplies over difficult terrain during the Falklands War, and take part in

RIGHT: A County forward control tractor with a Thompson dumper trailer. Designated 1164, this is possibly a pre-production prototype on trial.

LEFT:
A County FC 1174 fills up with fuel prior to starting work.

LEFT:
Built in 1977 for Hong Kong, a County 764 fitted with a HAP/2 crane by T. H. White of Devizes.

RIGHT:
A County 1184 TW on trial with an experimental Dowdeswell push-pull plough.

BELOW:
County 1474, powered by the Ford TW20 153 hp engine, working a Wilder pressure harrow in Hampshire, 1979.

RIGHT:
Flagship of the County range, the 188 hp 1884, introduced in 1980.

LEFT:
A pair of County 1884 tractors belonging to Wallasea Farms, Essex, photographed in 1984.

BELOW:
Two-wheel drive export model County 1162, fitted with the 120 hp Ford TW10 engine.

BOTTOM:
A high clearance version of the 1162, the County 1162H was designed for sugar cane work.

the clearing up operations in the aftermath of the conflict. But export sales were still falling and the company faced substantial losses in 1981 and 1982. With the pound strong against the dollar, County's machines appeared expensive on the depressed worldwide market, and the company found it difficult to compete. County had invested a lot of money developing new products, and with a slump in sugar cane production hitting further sales, the company found itself short of capital and went into receivership in February 1983.

The company was bought by David Gittins, a County dealer from Ludlow in Shropshire, who specialised in selling rebuilt County tractors to South America and other overseas markets. County Tractors Limited was formed, still operating from Fleet, and was back in business only nine months after the receivers were called in.

The County range was rationalised - the unequal wheel four-wheel drive K tractors were dropped, but the company continued to build all the equal-size four-wheel drive machines for both the domestic and export markets. Both the high clearance tractors and the Transit 4 x 4 remained in production.

County Tractors Limited was taken over in 1987 by the Benson Group, already owners of Bamfords and Teleshift and importers of Valmet tractors. The company was re-located to Knighton in Powys, and continued to produce a much reduced range of tractors, concentrating on the 774, 974 and 1184 models, as well as the Hi-Drive kit and the 4 x 4 transit. Towards the end of production, a

RIGHT: The Jubilee model County 1174, finished in gold for the company's fiftieth anniversary in 1979, and used for promotional work.

number of 1184 tractors were actually finished in Benson's red colour scheme. Of the other models, only one 1884, sold to British Steel at Birmingham, and three 1474 tractors were built at Knighton.

The company had problems finding suitable cabs to meet with UK legislation, and only about 150 tractors were built at Knighton before production finally came to an end at the beginning of 1990. County continued with the manufacture of the 4 x 4 Transits and the parts side of the business, supplying spares to an estimated 40,000 tractors working worldwide.

RIGHT: After the Jubilee model County 1174 was sold, the customer insisted that it was painted red and white. Today, it is owned by County dealers, A. T. Osborne Limited of Hampshire, and will be restored to its original gold livery.

LEFT: County 1474 at work with a Dowdeswell plough.

BELOW: County 1184 TW.

But the story does not quite finish there. SEM Engineering Limited, of Basildon, Essex, having acquired the rights to use the County name from Bensons, launched the County HSH 140 in 1994. A two-wheel drive, high-speed haulage tractor, developed for sugar cane plantations in Africa and capable of pulling 30 ton payloads at speeds of up to 30 mph, the HSH 140 was based on the New Holland Ford 8340 and powered by the six-cylinder turbocharged Powerstar engine developing 140 hp. Several machines are already working successfully in South Africa and Kenya.

SEM's engineering director, Eric May, has also redesigned the County four-wheel drive system to accommodate the Ford 40 Series skid unit, enabling County Tractors Limited to announce two new models of four-wheel

LEFT: A high capacity muck-spreader based on a forward control County FC 1184 TW.

RIGHT:
Prototype County 764 test unit fitted with a Perkins engine.

RIGHT:
County 764P export model fitted with a Perkins A4.248 engine.

drive tractors, initially aimed at the export market. To be built with basic eight-speed gearboxes, the 85 hp 764 (40) would be fitted with the four-cylinder Ford 6640 engine, while the six-cylinder 1164 (40) model would have the turbocharged 8340 power unit rated at 135 hp.

Work on the new County four-wheel drive tractors commenced at SEM's Basildon factory in late 1994. County Tractors had received an order for eight tractors for the Falklands, St. Helena and Canada, the first units to be ready for shipment in February 1995. Sadly, SEM Engineering went into receivership on 13 February 1995, and production of the new models was temporarily held up.

County Tractors Limited, having sold its parts business to Thama Holdings in the Midlands, had left itself free to concentrate on the 4 x 4 Transits, and the tractors and production of the new machines were transferred from Basildon to its own factory at Knighton, where the first units were finished during April 1995. The initial order, now extending to ten tractors, included the 75 hp 744-40 and 125 hp 116440 models without cabs and 85 hp 1184-40 tractors with cabs, and was completed in July 1995.

In May 1995, County released details of a range of four models, available with or without cabs, based on the Ford 40 Series tractors. Apart from the 75 hp 744-40 and 754-40 tractors, and the 125 hp 1164-40 and 1184-40 machines, County are also offering the 764-40 and 774-40 using the 85 hp Ford 6640 skid unit, and the 110 hp 784-40 and 794-40 tractors based on the six-cylinder 8240. Initially aimed at the export market, the new tractors will be fitted with basic eight-speed gearboxes.

At the time of writing, there is talk of the possibility of future County models with revised specification being introduced for the UK market. The company are looking into incorporating the Ford 12 x 12 Synchroshift and 16 x 16 Electroshift transmissions in its tractors, and are discussing with New Holland the prospect of gaining approval to use the Ford cab in the UK. Perhaps, before the first County tractor becomes fifty years old, we shall see new tractors bearing that name back on British farms.

BELOW: County 7600-Four specially adapted for mining use. The exhaust is fitted with a flash eliminator for working in explosive conditions

RIGHT:
Export model County 1164P fitted with a Perkins 6.354 engine.

RIGHT:
This 764 was one of two prototype County tractors fitted with the forward control modular cab in 1980, the other machine being based on the 1184. Both went to the Forestry Commission and no more were built.

LEFT:
This Dando drilling rig, owned by Balfour Beatty, is one of the many applications based on the County FC 1184 TW.

BELOW:
A Teg 750 timber forwarder working in Norway. Based on a County 754, it was fitted with a County Brockhouse 11F torque converter transmission.

RIGHT:
The Mersey Shunter converted from a County 1164 by Warrington Tractors. Fitted with a torque converter transmission, the tractor was designed for shunting railway wagons.

RIGHT:
The County Tractruk bulk spreader based on the Ford D series lorry.

LEFT:
One of 31 fire tender chassis assembled for Unipower by County at Fleet.

BELOW:
Believed to be the last County 974 built. Adapted by A. T. Osborne for the Forestry Commission, it is fitted with a special fork for pushing up and burning brash.

RIGHT: County Bulldog aircraft tug developed in association with Hallam Engineering of Leicester. Based on the County 764 and fitted with a Brockhouse torque converter.

RIGHT: County aircraft tug based on the 1184 tractor.

LEFT:
The County HSH 140 high speed haulage tractor built by SEM Engineering.

BELOW:
One of the new County tractors built in April 1995. This 125 hp 1184-40 was based on the Ford 8340, and sold to the Falklands.

Chapter 3

Doe

Doe Triple D tractor with a Doe eight-furrow semi-mounted plough in February 1964.

Today, Ernest Doe and Sons Limited is one of the largest agricultural and construction machinery dealers in East Anglia. However, for many years the company, based at Ulting, near Maldon in Essex, was involved in the manufacture of implements and machinery as and when the need arose. Demand for a large and powerful 4-wd machine to plough the heavy Essex clay soil resulted in the product for which the company is best known - the Doe Triple D tractor.

The history of Ernest Doe and Sons can be traced back to the nineteenth century. Ernest Doe, the firm's founder, was born in 1876, the middle son of Mr Charles Doe, a miller from Terling Mill. In 1893, after a short time spent in the mill, Ernest, aged seventeen, was sent to work for a local blacksmith, Mr Wood. After Ernest had completed his five year apprenticeship in 1898, the business was bought by his father from Mr Wood who was retiring and selling up. Doe's present depot at Ulting stands on the site of the original blacksmith's shop, affectionately known locally as 'Doe's Corner'.

Ernest Doe soon built up a high reputation for hard work and honest dealing and established good connections with the local farmers, including Mr James Havis, whose daughter he married in 1901. He and his wife had three sons - Ernest Charles, who followed his father into the blacksmith business; Hugh, who became a farmer; and Herbert Walter, who started a contracting sideline.

BELOW: The blacksmith's shop at Ulting in 1908. Ernest Doe stands on the extreme right of the group.

LEFT:
Ernest Doe Senior (left) pictured in the works foundry in 1958. This is now the site of the sales office.

The business quickly prospered under Ernest Charles Doe. He was keenly interested in tractors, and felt that they were the way forward for the future. Through the 1920s and 1930s he attended tractor trials and obtained agencies where he could. By 1937 the company advertised itself as 'The Tractor Specialists' and held agencies for Case, Allis Chalmers, Fowler, Fordson, and Ransomes tractors.

In 1941, Ernest Doe and Sons sold a total of 240 new tractors - 94 Fordson, 8 David Brown, 62 Allis Chalmers, 51 Case, 12 Caterpillar, 12 Cletrac and 1 Oliver 80. That same year, the company sent fifty old cross-motor Case tractors of 1920s vintage to be scrapped for the war effort.

LEFT:
Mr. Ernest Charles Doe.

To save on petrol during the war, it was suggested that Essex be divided into sections for each established agricultural engineer to operate in. This would have seriously affected Doe's business and lost them many of the customers they had outside their own area. To counter this, the company decided to open more

BELOW:
An advertisement for Ernest Doe and Sons' stand at the Essex Show in 1937.

ABOVE:
A new Case LA pictured in Doe's yard in 1941. In the background are Peterbro, Alldays, IHC Mogul and Fiat tractors waiting to be scrapped for the war effort.

ABOVE RIGHT:
A 1943 Doe conversion of a Fordson tractor into a mobile electric welder. The Murex electric welder mounted over the fuel tank was driven off a vee-belt taken from the tractor's adapted pulley.

depots, and legend has it that Ernest Charles Doe bought three branches on one Saturday morning.

After the war, Ernest Charles' son, Alan Ernest Doe, joined the company. The business was formed into a limited company with Ernest Charles, Herbert and Alan as directors. Ernest Doe Senior retired but continued to take an active interest in the business.

The company started to manufacture machinery both to compensate for the post-war shortage of implements, and to provide specialised machines for particular uses. A number of the popular Allis Chalmers Model B tractors were altered to low clearance by turning the final-drives upside down, and also often fitted with streamlined tin-work and

RIGHT:
Doe's high-clearance conversion of the Allis Chalmers Model B tractor towers over another Allis converted for orchard use.

LEFT:
The 1948 Wild Oat and Weed Seed Collector prototype.

narrowed for orchard work. A special high-clearance version of the tractor was made for working among fruit trees, the first being ordered by Seabrooks for spraying blackcurrant bushes.

In 1948, the company demonstrated its new Wild Oat and Weed Seed Collector. Mounted on a Fordson E27N Major, this remarkable contraption, known around the works as the 'V1' after Hitler's secret rocket weapons, was designed to collect wild oats, white clover and weed seeds off the ground. The machine was built by Doe for one of its customers, Mr A. H. Gardiner, who co-operated in the design. Basically a glorified vacuum cleaner, it worked by a series of spring tines agitating the surface of the ground and loosening the

BELOW:
The Tayler-Doe silage harvester pictured at the Royal Show where it was awarded a silver medal in 1950. Ernest Charles Doe (third from right) is standing next to the inventor, R.G. Tayler of Strutt and Parker Farms.

manager for Strutt and Parker Farms Limited, to manufacture a silage and green crop harvester based on a Massey Harris 726 combine harvester. All the threshing mechanism and grain collection parts of the combine were removed, leaving a basic chassis with the engine and drive system intact. The header was retained and fitted with a Reynolds pick-up reel and serrated knife to cope with the increased moisture of the crop. From the auger, the green crop material was fed straight up an elevator into a trailer towed behind the machine. Although designed originally to cut lucerne, the machine proved capable of dealing with most green crops, working at speeds of up to eight miles an hour and harvesting over 100 tons a day. With Massey Harris approval, several Taylor-Doe harvesters were sold throughout Essex and the West Country, and the machine won a silver medal at the 1950 Royal Show.

Ernest Doe and Sons Limited's business increased through the 1950s. They were supplying about 150 Massey Harris combines a year, and had fifteen service vans on the

ABOVE: Doe low-clearance orchard conversion of the Allis Chalmers Model B tractor.

RIGHT: Early Doe Triple D on industrial wheel centres. This tractor still has the early gear-change linkage.

FACING PAGE TOP: The early rod-linkage arrangement used to operate the high/low gear lever on the front unit of the Doe tractor.

FACING PAGE BOTTOM: The 1958 Doe's Dual Power tractor.

seeds which were drawn up a large tube, the suction provided by a fan running at 2,500 rpm driven from the belt pulley on the tractor. The seeds were then blown into a covered trailer towed behind the tractor. The tractor did not really have enough horsepower to provide adequate suction for the system, and the machine was not a great success.

The company entered into a more successful joint venture with Mr R. G. Tayler, area

road. In 1956, the company became exclusively Ford tractor dealers, beginning the long association with this make that continues to this day, although admittedly the tractors are now sold under the New Holland name. The company also entered the export market, sending a hundred Fordson E27Ns a month to Australia and New Zealand in conjunction with Wrights of Dereham.

Another venture at this time was the Doemobile, a universal 360 degree hydraulic excavator brought in from Sweden. Doe fitted it with a Ford power unit and modified it to meet UK regulations. Built on pneumatic tyres, it had a road speed of 25 mph, and could be used as a dredger, grab, crane or face shovel.

The story of the Doe Triple D tractor starts in 1957. Faced with farming many acres of heavy Essex clay, George Pryor of Navestock was looking for a tractor with the power and capability to work large cultivation equipment under difficult conditions. Believing crawlers to be unsuitable for his purpose, and there being little else available on the market, he set about building a tractor suited to his needs in his own farm workshop.

provide a four-wheel drive machine, with the two engines combining to give over 100 hp. The Fordson was used it was one of the few tractors with suitable clearance under the sump to allow for the conversion.

Ernest Charles Doe, keenly aware of the machine's potential, entered into an agreement with Mr Pryor which gave Doe manufacturing rights to the tractor. George Pryor built up a second machine, minus engines,

TOP: The Doe's Dual Power tractor, based on the Fordson Power Major, working in some sticky conditions.

ABOVE: A Dual Power tractor, fitted with Wright large-diameter steel wheels, works a seven-tine cultivator.

Initial developments consisted of using two standard Fordson Major tractors joined together by a beam. A driver was stll needed on each tractor, and the forward tractor still retained its front axle. While the system worked, it was far from satisfactory, and it was found that 20 per cent of tractive effort was lost just in pushing the front axle along.

George Pryor developed his ideas further, and the result was two Fordson Diesel Major tractors with the front axles removed, joined together and articulated in the middle, to which was sent to Doe. After fitting it with two spare Fordson engines, some of a number the company had left over from converting Howard trench diggers to Perkins power, Doe carried out a few further modifications of their own, and within three weeks, the first machine was ready for trials.

George Pryor's own original machine ran for ten years without problem before being dismantled. He subsequently bought two more machines off Doe, and still owns a Triple D which is destined for restoration.

LEFT: Dual Power tractor working with a Doe five-furrow mounted plough.

The first Doe tractor produced for sale was based on the Fordson Power Major, and called Doe's Dual Power. The two tractors were joined together by a specially built turntable 2 ft in diameter which was attached to the rear of the front unit. This in turn supported the front of the rear tractor unit, which was further strengthened by a sub-frame assembly. A longitudinal trunnion fabricated out of steel tube and mounted under the turntable allowed for lateral movement between the two tractors.

At over 19 ft long, the tractor appeared cumbersome and ungainly, and yet was surprisingly manoeuvrable. Efficient hydraulic steering was effected by two pairs of two rams working in tandem either side of the turntable. A compensating ram operated a Vickers control valve. The unit could be turned through almost 90 degrees, giving the machine a turning circle of only 21 ft. Power for the steering assembly was tapped from the standard hydraulic pump in the front unit, which was otherwise redundant with all

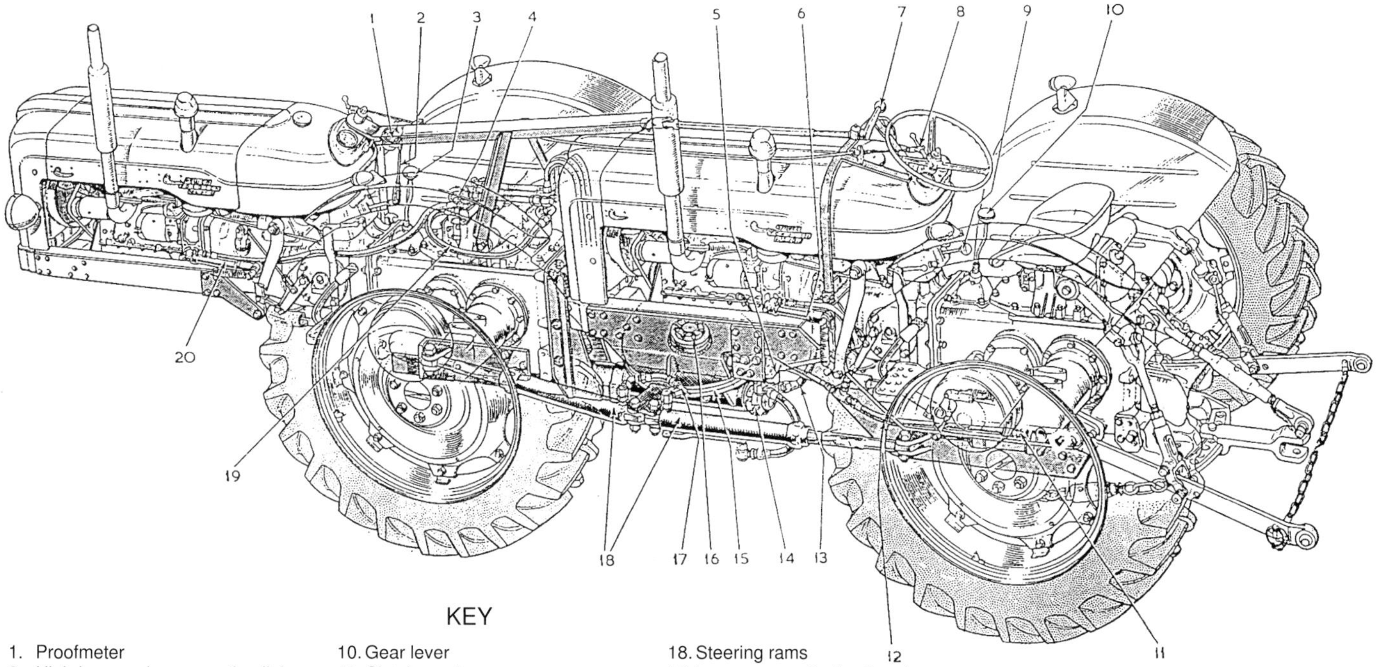

BELOW: Doe's Dual Power tractor. (Farmer & Stockbreeder *copyright*)

KEY

1. Proofmeter
2. High-low gearbox operating link
3. Gear lever
4. Steering hydraulic feed system
5. Steering control valve
6. Clutch slave ram
7. High-low operating lever (forward tractor)
8. Governor
9. High-low lever (rear tractor)
10. Gear lever
11. Clutch master ram
12. Hydraulic feed to clutch slave rams
13. Hydraulic feed to right-hand steering rams
14. Lateral pivots
15. Turn-table
16. Turn-table adjusting nut
17. Ram oil transfer pipe
18. Steering rams
19. Low-pressure hydraulic return
20. Clutch operating ram

RIGHT:
Demonstrator John Kent at the wheel of an early Doe Triple D based on the Power Major.

RIGHT:
A Power Major Triple D. Note that the primary gear lever is still in place on the front tractor unit. This was replaced by hydraulic slave rams on later machines

RIGHT:
The Power Major-based first version of the Triple D with Doe's twin leg subsoiler attached.

hydraulic requirements for implements being taken off the rear unit. The idea behind this was to spread the load on the engines equally between the two units.

The tractor was designed so that the driver seated on the rear unit could have control over both units. In reality he could only operate a few functions on the front unit of these early machines. Clutch operation was synchronised through the pedal on the rear unit using a hydraulic master cylinder and slave rams. The throttles were connected by

ABOVE:
The demonstration team. At the wheel, left to right: Dick Partridge who built the ploughs; Ernest Doe; John Kent, demonstrator.

BELOW:
The Doe Triple D - early version based on the Power Major. (Farm Mechanization *copyright)*

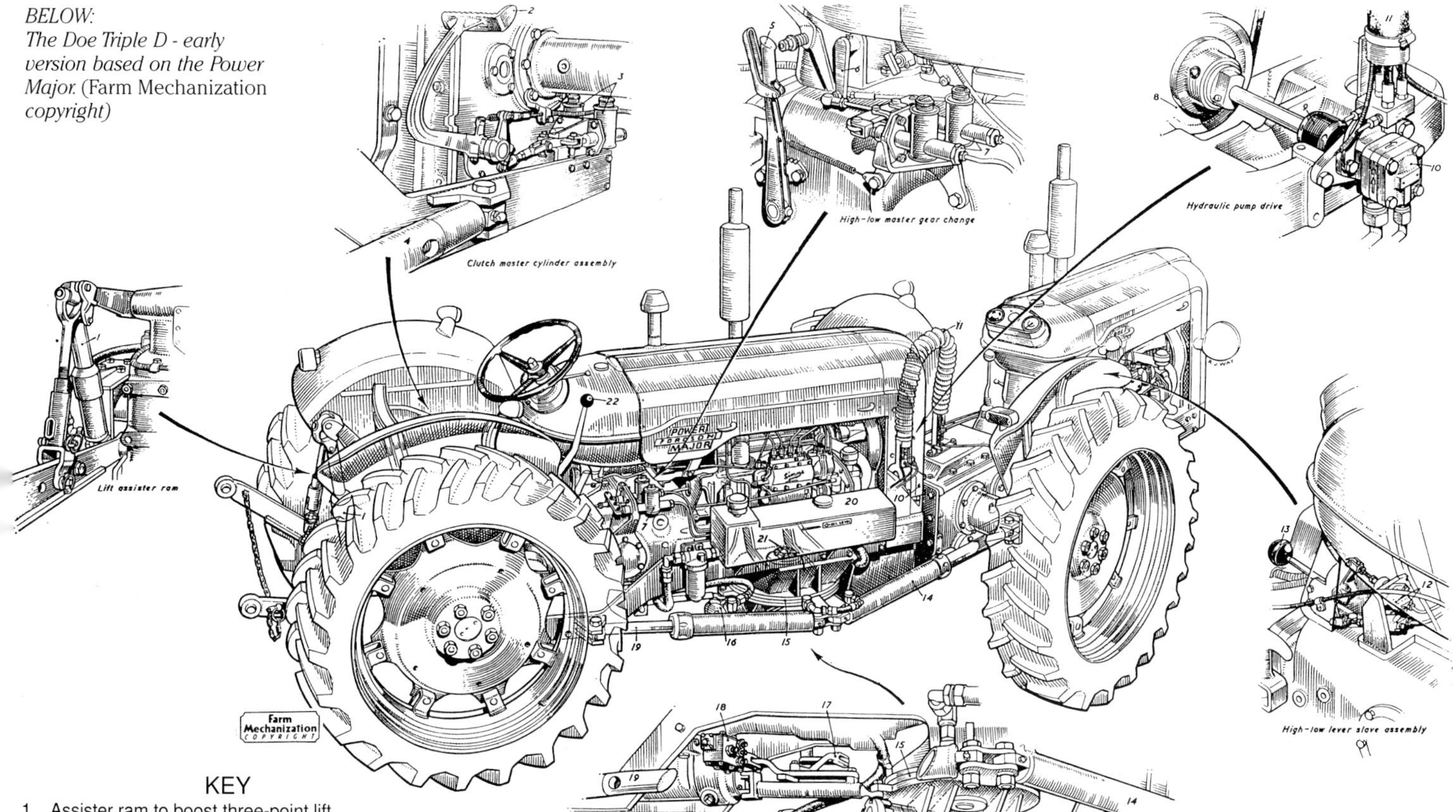

KEY

1. Assister ram to boost three-point lift
2. Master clutch pedal
3. Clutch hydraulic cylinders
4. Rear of left-hand pair of steering rams
5. Brake lever
6. Master high-low gear lever crank
7. High-low hydraulic cylinders
8. Fan belt pulley on rear engine
9. Hydraulic pump drive flexible coupling
10. Hydraulic pump for power steering
11. Conduit carrying clutch, high-low, throttle and pump cut-off controls to front unit
12. Hydraulic slave cylinders for remote control of front high-low lever
13. Front high-low lever
14. Front of right-hand pair of steering rams
15. Turntable assembly
16. Pivot shaft connecting front unit to turntable
17. Linkage between steering wheel and steering hydraulic control valve
18. Steering hydraulic control valve
19. Rear steering
20. Hydraulic steering oil-supply tank
21. Turntable pivot shaft
22. Throttle control lever coupled to both engines

RIGHT: Improved version of the Doe Triple D tractor based on the Super Major. The driver was now able to select all the gears on the front unit from his seat on the rear unit.

BELOW: The Super Major based version of the Triple D fitted with a Fritzmeyer cab. The plough is a Doe four-furrow reversible.

a long heavy-duty cable which alone cost over £100. A solenoid-type starter for starting the front unit from the rear unit was an optional extra.

The only control the driver had over gear selection for the forward tractor was through a system of link-rods running across the rear tractor bonnet which allowed him to operate the high/low gear lever on the front unit. With this in neutral, the driver walked round to the front unit and manually selected the gear he required to work in. On returning to his seat, he selected the same gear on the rear unit, used the

RIGHT: This photograph, taken at the rear of the Doe works, shows the degree of articulation that could be achieved with the Triple D tractor. Dick Partridge is at the wheel.

ABOVE:
Extolling the virtues of the the Doe Triple D tractor to a prospective customer. Note the special narrow cage-wheels fitted to the front wheels to allow turning.

LEFT:
1964 Doe Triple D, based on the New Performance Super Major tractors, tows a NIAE plastic pipe drainage machine. Only six tractors were fitted with the shield under the hull of the front unit as shown on this tractor.

sold by Doe went to Lordsfields Farms, West Suffolk, in the wet autumn of 1958, and was the only tractor able to continue ploughing in the poor conditions.

Only six of this early model Doe's Dual Power tractor were built. The tractor fell foul of the Motor Taxation Department which decreed that if the machine was to be classified and taxed as one vehicle then all the auxiliaries such as the hydraulics and power steering pump had to be mounted on the rear unit. Health and Safety officials were also unhappy about the method of selecting neutral on the front unit by use of the linkage arrangement which they believed to be unsafe.

ABOVE: Because of their export achievements with the Triple D tractor, Doe were invited to take part in the Lord Mayor's Show in London. The Doe Triple D, with workshop manager, Dick Sterk, at the wheel, is seen pulling the Milk Marketing Board float.

RIGHT AND OPPOSITE TOP: Doe Triple D tractors were exported all over the world. This example, being transported to the docks by the company's Ford Thames Trader lorry, is on route to the Leipzig Show in Germany

high/low gear selector linkage to take the front tractor out of neutral, and off he went. Reversing on headlands was achieved by placing the high/low gear lever of the front unit in neutral and using the reverse gear of the back tractor only. Not an altogether satisfactory arrangement, but it worked.

The tractor performed better than expected. Optimum results were obtained with the front engine set to run at a marginally higher speed than the back unit, so that the front tractor 'pulled' the rear unit. With almost perfect weight distribution, it coped admirably in all weathers. The first tractor

To go some way towards meeting these regulations, in 1959 the company brought out an improved version of the tractor known as the Doe Dual Drive, or Triple D, still based on the Fordson Power Major. To satisfy the taxation rulings, the hydraulic power for the steering was taken from a pump driven off the end of the crankshaft of the rear tractor unit. Selection of the high/low gear for the front unit was later changed with a hydraulic master cylinder and slave assembly rather

LEFT:
The 1964 Moscow Show. President Khrushchev, in the white hat (centre right), inspects a Doe Triple D tractor, the bonnet of which can just be seen in the foreground.

ABOVE: Doe heavy-duty toolbar fitted with twin subsoiler legs.

than the rod-linkage arrangement. An assistor ram was fitted to boost the three-point lift, and the tractor came out in Doe's distinctive colour scheme of blue with orange bonnets. List price for the machine was £1,950 in standard specification.

With the introduction of Ford's Super Major tractor in 1960, the Triple D was improved still further. The driver was now able to select all the gears for the front unit from the back tractor through a system of master and slave cylinders working on the front gear selector, shortened and held captive in an enclosed box. A lever fitted to the right of the driver was pushed forward to select first gear for the front tractor. Pulling it back engaged second. With the lever in a central neutral position, pressing a foot-pedal moved the gear selector across the gate, thus allowing the lever to select third or reverse. With this lever back in neutral, the selector would automatically be pushed back by a spring to the first/second gear position. The differential lock on the front tractor could also be operated hydraulically from the rear unit. All the hydraulic pipework for the controls was enclosed in a flexible rubber tube passing between the two units.

The new Triple D won a silver medal at the 1960 Royal Show. Put to test on Lord Rayleigh's Farms near Chelmsford in August 1961 with a six-furrow plough, the tractor ploughed over thirty acres in twelve hours – an

BELOW: The seven-tine cultivator based on the heavy-duty toolbar

LEFT:
Doe's hydraulic folding toolbar fitted with two sets of Ransomes 9 ft disc harrows.

average of 2.5 acres an hour – and it used just over thirty-five gallons of diesel.

When the uprated New Performance Super Major came on the market in 1963, the Triple D based on it could boast 108 hp. Four-wheel drive coupled with the advantage of having a differential lock working on all four wheels made the Triple D a formidable machine in the 1960s when few other comparable tractors were available.

However, the Triple D did have several drawbacks. Driving a machine nearly 20 ft long that weighed over five tons on the road was no mean feat. Problems with wear on the compensating ram when the tractor was in a straight line tended to give somewhat erratic steering when travelling at any speed. Also, an inexperienced driver could easily get the two tractor units into two different gear ratios, or even one in forward and one in reverse gear at the same time, with disastrous and expensive results.

Because both the hydraulics and the power steering were all now fitted to the back tractor, the engine of the rear unit tended to work harder and wear out quicker. Further problems were encountered with movement in the top-covers fitted to the Super Major back-ends when using heavy implements. Doe got round this by fitting dowels between the top-covers and the main castings, and offering an optional heavy-duty linkage arrangement. The Triple D was originally

BELOW:
The hydraulic folding toolbar fitted with spring tine cultivators giving a working width of 20 ft.

RIGHT: 1962 Doe Triple D tractor. This machine was once owned by the author.

BELOW: A New Performance Super Major version of the Doe Triple D demonstrates its ability to turn through 90°

designed so the owner could turn it back into two tractors for the summer months. In practice few farmers bothered because of the work involved, and Doe only ever supplied one conversion kit of the parts necessary to do this.

Unfortunately production records for the Doe Triple D have been partly destroyed by fire and are incomplete. What records remain show that nearly 290 Dual Power and Triple D tractors were built between October 1958 and October 1964. The tractor was demonstrated around the world, and twenty-eight were exported to places including Europe, Scandinavia, Africa, Australia, North and South America. In the USA, the machine was known as the Double-T and was sold through Tandem Tractors Inc. of Selby, South Dakota. A Triple D was exhibited at the Moscow Show in 1964, and three were sold to the USSR. Doe also designed a road-roller based on the Triple D arrangement. The machine never went into production. Only one example was built, and is believed to be still in existence.

Ernest Doe & Sons Limited were not content with just building the Triple D tractor. As very few large implements were available at the time, the company manufactured a complete range of ploughs and cultivation equipment for use with the tractor. The company's range of implements included a heavy-duty toolbar frame which could be fitted with seven cultivator tines, two subsoiler blades, or a mole drainer attachment. Doe also built a toolbar for use with disc harrows or spring-tine harrows covering up to 20 ft, which could be hydraulically folded to allow passage through narrow gateways and transport up the road. The range was eventually extended to include heavy-duty disc harrows and ripper ploughs.

Originally, Doe only offered a conventional five-furrow mounted plough for use with the Triple D tractor, priced at £235 in 1960, but soon four- and six-furrow versions were added to the range. In 1961, the company brought out its first reversible plough. Built

ABOVE: Doe five-furrow semi-mounted plough with steerable headstock reduced to four furrows by the removal of the back body.

LEFT: Triple D at work with the Doe five-furrow mounted plough. The tractor is equipped with the optional heavy-duty linkage which cost £169 extra.

as a four-furrow with a tubular frame, it proved difficult to set and only twenty were built. The following year the company brought out a new reversible plough with a normal bar-frame which was more successful. Three-, four- and five-furrow versions were eventually produced.

Doe's most popular plough was a universal semi-mounted steerable plough, convertible to four or six furrows by the removal or addition of a leg and body. The company also built an impressive eight-furrow version of the semi-mounted plough. Normally Ransomes plough bodies were used by Doe in the manufacture of their ploughs, but parts supply became difficult, sometimes taking up to four months, so the company turned to Lemken who built parts to Doe's own specifications. Doe offered a choice of three plough bodies. The DHS (Doe High Speed) 1 was a general purpose body similar to a Ransomes YL. The DHS 3 was a full semi-digger body, while the DHS 2 was a cross between the two.

The customers for Doe's range of ploughs

ABOVE:
Doe's early tubular beam four-furrow reversible plough in 1961. Only twenty were built.

RIGHT:
The later Doe bar-frame four-furrow reversible plough.

and implements were not just Triple D tractor users. As other makes of larger tractors became available or were imported, farmers turned to Doe to supply them with the bigger tackle they needed. Doe even manufactured a tool carrier so that the implements could be used behind crawler tractors.

With the introduction of the new range of Ford tractors from Basildon in late 1964, Doe redesigned their tandem tractor around two Ford 5000 power units. The resulting new machine was known as the Doe 130, the '130' figure designating the combined power of the two 65 hp 5000 units. The basic configuration remained unchanged, although the sub-frame and trunnion were strengthened to cope with the increased stresses of a more powerful machine, and the first few machines were fitted with two steering rams rather than four. An electric solenoid was added to select fifth gear on the front unit. The Doe 130 was listed at £2,975 in August 1965. The performance of the Doe 130 was initially hampered by the wide ratio gearbox fitted to the early Ford 5000s, although this improved when Ford revamped the

ABOVE: Demonstration of the impressive Doe eight-furrow semi-mounted plough.

LEFT: Triple D working with the Doe tool carrier and four-furrow reversible plough.

speed ranges. Just over seventy Doe 130s were built, and unfortunately only a few still exist as many were broken up for the export market.

The Doe 130 evolved into the Doe 150 when the new Ford Force range arrived in 1968, and the 5000 was uprated to 75hp and priced at £4,200. Only a handful were made. Doe's tandem tractor had nearly reached the end of the line. Demand for the concept was falling as many other more conventional high-powered four-wheel tractors were coming on to the market. Also Doe were not making much money out of the machine at this time, and preferred to concentrate on the retail side of the business, which was booming. In fact, on the opening day of the 1968 Royal Show, Ernest Charles Doe placed an order for ninety Ford Force tractors with Ford Tractor Operations. At the time this was the largest single order ever received by Ford from their UK agricultural dealer network, and was worth over £90,000.

The introduction of safety cab laws finally brought the Doe tandem tractor production to an end. The Ford safety cab was tested for the weight of one tractor, but not the combined weight of two machines as used in the Doe unit. One Doe tractor was built out of two Ford 7000 units and was designated

LEFT:
A Doe 130, fitted with oversize tyres, still at work in 1983.

FACING PAGE TOP:
The Doe 130 tractor.

FACING PAGE BOTTOM:
One of the earliest Doe 130 tractors built, supplied to Orsett Estate, Essex, in June 1965. It completed over 2,000 hours in its first two seasons work with no major breakdowns.

LEFT:
Photographed in Germany, one of the Doe 130 tractors that was exported to Europe.

Doe 180, and a few original 130s were returned to the works to be uprated to later specifications. One or two of these were fitted with 7000 units at the rear and 5000 units on the front to compensate for the power loss experienced with the rear engine.

The Triple D and its sister machines were not the only tandem tractors to reach the market. In fact the concept of joining two tractors together to produce a machine of increased power and tractive ability goes back long before the Doe.

The use of tandem tractors became fairly widespread in the USA following pioneering work by the Iowa Agricultural Experimental Station in 1955 who coupled togther a Ford 8N and a Ford 860 tractor. The system was developed further in conjunction with Michigan State University Agricultural Engineering Department, and in 1957 they exhibited a more sophisticated machine based on two Ford 860 tractors.

Experimentation on the tandem tractor concept appears to have taken place all over the world. In 1953 two Field Marshall tractors were used in tandem in Australia. A more successful arrangement on two Case 832C tractors equipped with torque converters was produced by Plant & Plant of Queensland. In France, Someca joined two of their 45hp tractors together in tandem, while A. H. Steenberger of the Netherlands produced a machine based on two International 436

BELOW:
Steering ram arrangement on the Doe 130.

RIGHT:
The 130 bhp Doe 130 tractor at work.

OPPOSITE TOP:
A line-up of eight Doe 130 tractors destined for export to Nigeria, in 1965, for sugar cane work. All the tractors had top-gear blanked off.

OPPOSITE BOTTOM:
A special export specification Doe 130 for the Australian market.

BELOW:
Industrial version of the Doe 130 supplied new to Fitzpatrick in 1965 for work on road building.

tractors. Even in Russia experiments were carried out on two Vladimir T28 tractors coupled in tandem. In Sweden, BM Volvo enjoyed limited success with a tandem tractor based on two of their 35hp three-cylinder 350 tractors, but production was eventually stopped due to lack of interest in the country.

The Doe Triple D had little competition in the UK. Berkshire farmer, Pat Saunders, developed a dual tractor kit which was marketed by Paramount Engineering of Coventry and allowed any two popular makes of tractor to be joined together but it never really caught on.

The gradual demise of the tandem tractor concept did not quite mark the end of Doe's tractor building days. In 1971 they built a 100hp two-wheel drive tractor based on the Ford 5000 and known as the Doe 5100. Aimed at the farmer who required more horsepower for such as silage work, it was fitted with the Ford 2704E six-cylinder industrial engine. Heavy-duty sub-frame castings to take the engine were imported from the specialist tractor builders, EVA of Belgium. The clutch assemblies were made up for Doe by County. Adverse reaction to the tractor from Ford led to the machine being dropped and only three

ABOVE:
One of two Doe 130 tractors purchased in July 1966 by E.A. Sheardown and Co. Ltd. of Grantham, Lincolnshire. By the following January, the tractors had each done between 1,700 and 1,800 hours' work on cultivating and pea vining.

RIGHT:
Sheardown's Doe 130 working a Doe 16 ft 15-tine ripper plough at a rate of 80 acres per 24 hours at peak periods.

RIGHT:
1957 Michigan State University tandem tractor based on two Ford 860 units.

or four were built.

Doe manufactured little else after this time as the retail and distribution side of the business spread throughout East Anglia serving the farming, construction and horticultural industries. However, the Doe story is not complete without a mention of the Colchester Tractors side of the business.

ABOVE:
A 1966 Doe 130 tractor at the end of its working days.

Looking to open new branches, Ernest Charles Doe bought 79 Hythe Hill, Colchester from coal merchants, Thomas Moy, in 1943 and operated the premises as an Ernest Doe and Sons branch until 1947. When the local Fordson dealer, Willetts, went into liquidation in 1946, the Fordson franchise for the Colchester area became available. The Ford Motor Company would only give Ernest Doe and Sons the franchise on the condition that they gave up selling other makes of tractor. This was unacceptable to Doe, but a compromise was reached in that Colchester Tractors Limited was formed on 16 May 1947, to operate from the Hythe Hill premises. The directors of the new company were Ernest Charles Doe and Herbert Walter Doe with Stephen Woodridge as managing director. In 1952, Ernest Charles' son, Alan, took over as managing director, a position he held until 1957 when he was succeeded by Owen Dunt.

During Dunt's time as managing director of Colchester Tractors, a subsidiary company, Colchester Tillage Limited, was set up. Colchester Tillage's main concern was the import of Lemken ploughs and cultivation equipment, but the company also imported Tico cranes which it fitted to Ford tractors.

Early Colchester Tillage crane conversions

LEFT:
A later conversion of the Doe tractor, uprated by having later Ford Force 5000 (front) and 7000 (rear) power units fitted.

BM350
BM350
MIAK

LEFT:
A Muir-Hill 101 tractor comes to the aid of a stranded BM Volvo tandem tractor.

OPPOSITE TOP AND BOTTOM:
A BM Volvo tandem tractor working in Sweden, with two ploughs also coupled in tandem.

BELOW:
Colchester Tillage Cotil CT74 crane based on a Ford 5000.

were built on Power Major and Super Major tractors. In 1965, the Cotil CT74 one-ton crane based on the Ford 5000 tractor went into production. By the 1970s the range included 2 ton cranes also built on the 5000, and larger 2.5 ton capacity machines based on County tractors.

With the sale of a significant part of the Tillage business to Lemken UK in 1993, it was felt that the cost of Colchester Tractors remaining independent was too high. The Colchester group of companies was still controlled by the Doe family. So in November 1993, Colchester Tractors Limited was purchased by Ernest Doe and Sons Limited who also bought the trade and trading assets of Colchester Tillage Limited. The business is now reorganised as a branch of Ernest Doe and Sons - back in Colchester after nearly fifty years. Colchester Tillage operates as a trading division of Ernest Doe and Sons Limited, importing Cormach lorry cranes along with other parts for past Tillage franchises.

At the time of writing, Ernest Doe and Sons Limited is a very successful retail organisation operating twelve branches across East Anglia and employing nearly four hundred people. The company's turnover for 1994 was well in excess of £50 million. Alan E. Doe is now company chairman, and his son, Colin Ernest, great-grandson of the founder, is managing director, keeping the tradition of a family-run business alive.

RIGHT:
Cotil CT74 working in a warehouse.

RIGHT:
Cotil CT74 showing the crane in the folded transport position.

RIGHT:
Cotil CT74 at work on a building site

LEFT:
Cotil CT5004 crane based on a County 1124 tractor.

RIGHT:
A Cotil crane based on a County tractor. The tractor is fitted with a Wickham Poole hitch.

BELOW:
A Cotil CT4504 crane mounted on a County 1184 tractor.

CHAPTER 4

Chaseside & Northrop

Chaseside's Leyland Comet and semi-trailer unit, carrying an early Super-Loadmaster four-wheel drive loading shovel, waits outside the British Northrop works at Blackburn.

NORTHROP
CHASESIDE
CHASESIDE
ENGINEERING COMPANY LIMITED
STATION WORKS
TEL: HERTFORD 3113
LEYLAND
4489UR

The Chaseside story started in 1925, when George Hunt Jackson, a Ford distributor, produced the first British-built mechanical loading shovel as a means of using up his surplus stock of tractors.

Jackson, born in Britain where he served his apprenticeship, qualified as an engineer in Canada, returning to England at the start of the First World War. After the war, he set up the Chaseside Motor Company, agents for Ford motor vehicles and agricultural engineers selling Fordson tractors. The company name was derived from the location of its premises, off Enfield Chase, Middlesex.

Seeking to reduce stocks of Ford vehicles that were not selling as fast as he would have liked, Jackson found that converting the basic units into special-purpose machines generated new sales outlets. These early conversions included a Model T Ford truck with its load-capacity increased by the fitment of strengthened road-springs and a heavy-duty rear-axle, as well as shunting tractors and the loading shovel.

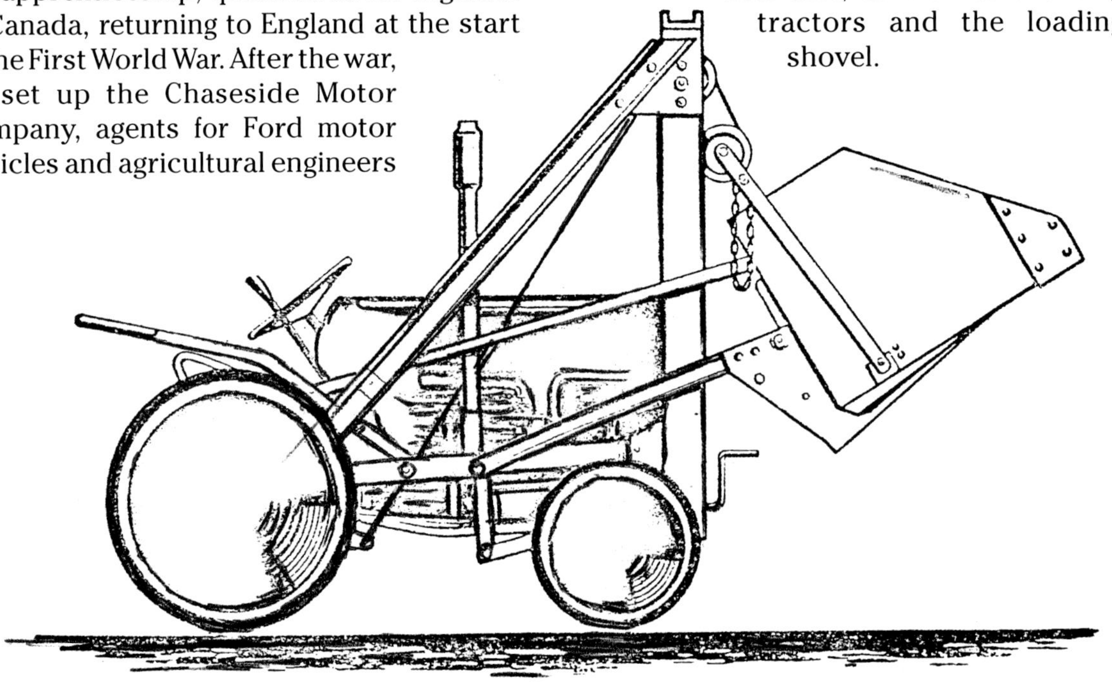

RIGHT: Artist's impression of the 1925 Chaseside loading shovel.

The original 1925 loading shovel was based on the Fordson Model F tractor. The ½ cu yd front shovel attachment was operated by a wire rope from a rear-mounted cable winch, which was chain driven from the tractor's belt pulley shaft. The boom support was fabricated out of steel channel with welded joints, and allowed overhead positioning of the rope sheaves. The shovel bucket was tipped and latched by gravity. Considerable interest was shown in what was the first wheeled loading shovel to be available in Britain, and Chaseside were soon taking orders for several machines.

The basic design of Chaseside shovels remained unchanged for the next thirty years. Chaseside remained faithful to the rope-operated system, and all their machines were built on Fordson tractor skid units. However, substantial improvements were made to the machine during its long production run. Hi-Lift and Lo-Lift versions with optional high or low lift side arms were intro-

BELOW: A 1938 Chaseside shunting tractor.

duced, as were road lighting and various other attachments. The jib became adjustable for height to enable the machine to work in confined areas. On the 1934 Hi-Lift shovel, the cable winch was moved to a mounting on the frame above the driver's head, a similar position to that used on the contemporary Muir-Hill shovels. This was a short-lived arrangement, and the winch drum was soon returned to its rear mounting, which remained a feature of all Chaseside rope-operated shovels.

ABOVE: A 1934 Chaseside Hi-Lift shovel. Lined up at the rear are some of the company's other early products, including commercial vehicle conversions.

The Chaseside loading shovel fast gained in popularity, as the only real alternative at the time was the ungainly and expensive steam shovel. The Chaseside was rugged and cheap to maintain, manoeuvrable and relatively easy to use. Requiring only one man to operate it, and loading 20 to 30 tons of material an hour, it was truly the machine that replaced the navvy's hand-shovel on building sites all over the country.

The company also built a mobile tractor crane based on the loading shovel. The crane used the same main frame and system of rope operation, but had a different jib arrangement. It was designed to lift 10 to 15 cwt to a maximum of 9 ft 6 in, although lift height was increased on later models.

LEFT: A 1934 Chaseside Mobile Tractor Crane, designed to lift 10 to 15 cwts to a height of 9 ft 6 in.

One of the strangest Chaseside products was an aeroplane built in 1933 by the sales manager, Mr Curtis, an avid aviation enthusiast. Naturally, it had a Ford power unit, using a standard 24 hp Model A engine, heavily modified to give a remarkable 66 hp. Built as a bi-plane, it would cruise at 85 mph. Plans were made to put the aeroplane in production, priced at £450. How many were actually built, or sold, is not known.

As demand for Chaseside products grew, George Jackson realised the need to expand. In 1937, he formed the Chaseside Engineering Company Limited, and bought land at the back of the Chaseside Motor Company as a site for the new engineering works. Up until this time, the manufacturing side of the business had not been run as a separate concern from the motor company. Jackson had two sons, William and Gordon. Eventually, William was put in charge of the Chaseside Motor Company, and Gordon became director and general manager of Chaseside Engineering.

Further improvements were carried out on the loading shovel between 1938 and 1945. A frame was extended around the driver, partly for protection against the lethal consequences of the wire cable breaking. Operators were advised to ensure the cables were kept in good order. The frame was fabricated from welded rolled steel sections, and the shovel boom and jib from electrically-welded steel. A friction-type winch with a double friction drum was fitted, driven by a Reynold duplex chain.

TOP:
A 1938 Chaseside Super-Lift shovel.

ABOVE:
A line-up of 1938 Chaseside mobile cranes.

RIGHT:
A 1942 Chaseside shovel, belonging to Fitzpatricks, working near St Paul's Cathedral in London

LEFT: A 1943 Chaseside Hi-Lift loading shovel.

The driver would raise the bucket by pushing down the winch lever on his right, until the desired height was achieved. The lever would then be released and the drum brake would automatically hold the bucket in the required position.

During the 1940s, the company looked to agriculture for further outlets for its machines, and built the Chaseside Hi-Lift Farmyard Manure Shovel. Chains were substituted for tie-rods on the top of the machine so that the jib could be lowered for working in barns or covered yards.

The Chaseside Engineering Company was soon looking to expand further, which was not possible at its present location. The company did not wish to move to a development area and lose much of its workforce, so a new factory was built only 15 miles away on a redundant part of the station yard off Gashouse Lane in Hertford. As it was on a direct railway line from Enfield, most of the workforce could commute by train. The new factory was known as Station Works. Another advantage of moving to Hertford was that the

BELOW: A Fordson N based Chaseside Hi-Lift loading shovel fitted with Rotaped tracks.

RIGHT:
The Chaseside loading shovel, fitted with Rotaped tracks, on trials.

RIGHT:
An experimental wartime Chaseside dumper based on a Fordson tractor skid unit fitted with Rotaped tracks.

company still had reasonably easy access to London Docks, important to a company that had steadily rising overseas sales. The Chaseside Motor Company also eventually opened a branch in Hertford, and emphasised the agricultural side of the business by claiming to offer farmers ' the complete mechanical farming service'.

ABOVE: Demonstrating the Chaseside experimental dumper on Rotaped tracks.

With the advantage of newly increased production capacity, the company took the opportunity to further increase its range of loading shovels, all based on the improved Fordson E27N Major that became available in 1945. The ½ cu yd machine became known as the Hi-Lift Light Excavator when the larger ¾ cu yd and 1 cu yd machines were introduced, along with the extra-high-reach Demon shovel. A range of attachments were also offered for the machines, such as an angle-dozer, a bulldozer blade and various special buckets including a slotted clay handling scoop. A one-ton crane version of the shovel was also built.

A new product, the Chaseside Hi-Speed Dumper, was also introduced in 1947, available in either 2½ or 3 cu yd form. The dumper was based on the Ford 7V truck, from

LEFT: A 1946 Chaseside ½ cu yd Hi-Lift Light Excavator, based on the Fordson E27N Major, working in a gravel pit. Note that the machine is fitted with twin wheels for extra traction.

which it took its V8 engine, transmission, front axle and steering. The commercial Ford rear axle, not strong enough to take the full weight required, was fitted as a solid axle with chain drive to the main axle.

Chaseside shunting tractors had been built in very small numbers since the 1930s, fitted with buffers, front and rear, for use in railway sidings. The company continued to offer the Chaseside Buffer tractor based on the Fordson E27N Major, and the later New Diesel Major into the 1950s.

Export sales for 1949 had increased by 117 per cent, and the company claimed that between five and six thousand of its machines were at work in forty-five different countries of the world by 1950. The North American market seemed to elude Chaseside's attempts to break into it. To try to rectify this, a Canadian subsidiary was set up, and in 1949, Gordon Jackson spent six weeks in Canada transporting a Chaseside loading shovel on a low-loader from Winnipeg to Vancouver in an attempt to encourage sales.

By the beginning of the 1950s, there was a certain amount of dissension among the board of directors over whether Chaseside

LEFT:
Largest of the Chaseside rope-operated shovels, the 1 cu yd machine handling coke at a gasworks in Hertford.

FACING PAGE TOP:
A Chaseside ½ cu yd Hi-Lift shovel fitted with a farmyard scoop loading old battery plates.

FACING PAGE BOTTOM:
A ¾ cu yd Chaseside Demon shovel loading a lorry with coke at St. Helens Gasworks in Lancashire.

LEFT:
A Chaseside Hi-Lift excavator backfilling with a 35° angledozer attachment.

should continue to build rope-operated loading shovels, which were becoming outdated. Company accountant, Fred Weatherill, left Chaseside in 1951 and set up his own company building hydraulic loading shovels, believing this to be the way forward for the future. His first machine, the Weatherill 2H, still based on a Fordson skid unit, sold over 1,000. This galvanised Chaseside into action and they started looking at the use of hydraulic power in their loading shovels, bringing out the Loadmaster range of hydraulically operated shovels in 1955.

RIGHT: A Chaseside rope-operated shovel, based on a Fordson E27N Major powered by a Perkins P6 engine, lies derelict in Northamptonshire

First of the Loadmasters was the LM 500, a ¾ cu yd shovel based on the Fordson Diesel Major tractor. A reversed version with front-wheel drive, known as the LM 600, was also introduced. In 1958, the LM 700, a 1 cu yd shovel with further improvements including self-levelling linkage, was introduced along with the LM 750 with increased discharge height, and the reverse drive 1 cu yd LM 800 and larger LM 900 shovels. Based on the Power Major, these machines had the option of a Brockhouse torque converter transmission.

In 1959, Chaseside introduced the four-

RIGHT: Rear view of the Chaseside shovel showing the winch position and ballast boxes.

RIGHT:
Fordson E27N Major-based Chaseside one ton crane laying a 21 ft long steel sewer pipe across the Leeds South main road.

RIGHT:
Chaseside 3 cu yd Hi-Speed dumper, based on Ford 7V components.

ABOVE:
A Chaseside 2½ cu yd Hi-Speed dumper is loaded by a Hi-Lift shovel fitted with a slotted clay handling scoop.

RIGHT:
Chaseside Buffer tractor based on the Fordson E27N Major.

LEFT:
A 1947 Chaseside shunting tractor moving railway trucks at the West India Docks.

BELOW:
A 1956 Chaseside Buffer tractor, based on the Fordson Diesel Major and priced at £815.

wheel drive LM 1000 loading shovel, a first departure for the company from designs based on Fordson agricultural skid units. Although fitted with a Ford six-cylinder engine, it was a completely new purpose-built machine with Kirkstall four-wheel drive axles. The company claimed it was the first British-built four-wheel drive loading shovel, but Bray's four-wheel drive BL 430 machine had been in production since 1956.

A totally new Perkins powered dumper also went into production, designed by George Liskein, Chaseside's Austrian-born designer. The prototype of this machine is unofficially credited with bending a Hertford County Council dustcart when its steering axle broke during road trials. Reputedly, the dustcart was hurried through Chaseside's workshop before any local official could become any wiser.

The cost of developing the hydraulic loading shovels had put a drain on Chaseside Engineering's resources and, in 1959, the company merged with British Northrop Limited of Blackburn in Lancashire, Gordon Jackson becoming the group managing director. British Northrop were long-established loom and weaving machinery manufacturers. J. A. S. Northrop was an Englishman

RIGHT: A Weatherill 22H hydraulic loader built in the early 1960s, based on the Fordson Super Major. The Hertfordshire-based company was started by Fred Weatherill, Chaseside's company accountant.

RIGHT: One of Chaseside's range of hydraulic loading shovels introduced between 1955 and 1958, the Loadmaster LM 700 was based on the Power Major skid-unit.

who developed a system for changing the bobbin automatically on rotary magazine looms in the States. Two British entrepreneurs, H. P. Griggs and William Livesley, saw the invention at work during a visit to the USA, and arranged to incorporate the concept into British looms. On returning to England, they started British Northrop in 1902, moving to the Blackburn site in 1907. During the Second World War, the company was involved in armament and machine tool manufacture, expanding its machine shops further in 1946. Although by the 1950s, Northrop still had a thriving export market to third-world countries, demand for their products had fallen in the UK, leaving them with considerable unused factory capacity including a grey-iron foundry, and so the merger with

ABOVE:
Loadmaster LM 700 at work outside the British Northrop factory in Blackburn.

LEFT:
The later Chaseside dumper, designed by Austrian George Liskein.

ABOVE: The Chaseside dumper is loaded by the Loadmaster LM 1000 shovel acting as a crane.

Chaseside was a useful diversification.

Main production of the Chaseside machines was moved to Blackburn, and from 1962 to 1964, a completely new range of four-wheel drive Super-Loadmaster (SLM) loading shovels were introduced, using Ford, Perkins or Leyland engines, and Clark or Allison torque converter and power shift transmissions. Top of the range, the SL 3000, was a 3 cu yd machine with a capacity of 10,000 lbs.

LEFT: Chaseside LM 800 front-wheel drive, rear-wheel steer, loading shovel, still based on a Fordson skid-unit.

The established range of two-wheel drive loading shovels continued to be built in Hertfordshire, and a new depot was set up at Church Street in Ware as a sales and service centre. The company also operated from here as southern counties concessionaires for the Italian Same tractors for a while from 1963.

One of the most interesting products to come out of the British Northrop / Chaseside merger was the Northrop tractor. Although bearing the Northrop name, it was essentially a Chaseside product, and an exciting, if ill-fated, divergence by the company into the agricultural market.

LEFT: Chaseside Super-Loadmaster SLM 2000 undergoing severe trials.

The Northrop tractor project was initiated in 1964 by David J. B. Brown, who had joined Chaseside as chief executive at Blackburn after a spell working in timber plantations in West Africa. D. J. B. Brown's experiences in the logging industry had convinced him that the market required a basic four-wheel drive machine, built around a conventional agricultural tractor, that would suit the needs of the average farmer and yet, at the same time, have the manoeuvrability and adequate ground clearance to meet the requirements of the forestry and sugar cane industries. Designer Frank Thornton drew up plans for

LEFT: Publicity shot of the prototype Northrop 5004 tractor.

RIGHT: Prototype Northrop 5004. The tractor is fitted with a standard eight-speed gearbox.

BELOW: Early 5004 Northrop, fitted with the Select-O-Speed gearbox, photographed at Blackburn. Note the prop-shaft drive to the front axle.

ABOVE:
Production model Northrop 5004 out on moorland above the Blackburn works.

the tractor, working to guidelines set by D. J. B. Brown. The new Ford 5000 skid unit was chosen as a basis for the machine, partly because it had the advantage of a parts and service network established almost worldwide.

The first prototype was ready by December 1964. Power from the standard Ford four-cylinder engine was transmitted via the normal eight-speed gearbox to a drop transfer box. Both axles were fabricated from Ford 5000 parts. The drive-steer front axle had planetary hub reduction gears, and its centrally mounted differential was driven through a propeller shaft from the transfer box. Several initial teething problems were encountered and, when the prototype was started up for the first time, its front and rear wheels were found to revolve in opposite directions.

The first Northrop tractor went to work on T. A. Saul Limited's farm in Bedfordshire in early 1965.A few months later, D. J. B. Brown left Chaseside to join Muir-Hill as managing director where he saw through the introduction of the Gloucester-based company's range of four-wheel drive tractors, similar in concept to the Northrop. Brown persuaded Chaseside's chief engineer, Eric King, to join him at Muir-Hill, and responsibility for the Northrop project was left in the hands of the newly appointed chief designer, Gilbert McIntosh, working with Frank Thornton.

BELOW:
Northrop 5004 at an agricultural show in the 1960s.

RIGHT:
The Northrop drive-steer front axle was fabricated from Ford 5000 rear axle parts.

The tractor went into limited production in the summer of 1965. Known as the Northrop 5004, it was priced at £1,980, and was offered with a choice of eight-speed or ten-speed Select-O-Speed gearboxes. A torque converter transmission was advertised, but no tractors were ever actually fitted with one.

A spin-off of the tractor project resulted in the Ford 5000 skid unit with axles similar to those used in the Northrop, but reverse drive, used as the basis for Chaseside's new Super-Loadmaster SL 800 loading shovel, with twin-disc torque converter transmission, introduced in May 1965. A Belgian company, Comet, also built a backhoe-loader based on the Northrop 5004 tractor.

Northrop Tractors Limited was set up, and operated from Church Street in Ware under sales director, J.E.T. Watson, assisted by Les Carter who acted as service department manager and demonstrator, although all the tractors were built at Blackburn, including all the necessary castings which were

RIGHT:
Northrop 5004 ploughing heavy clay with a Bamford four-furrow plough in Berkshire, 1967.

LEFT:
SL 800 loading shovel, introduced May 1965, and based on a reversed Northrop skid-unit.

BELOW:
Belgian-built Comet back-hoe loader based on the Northrop 5004.

made in-house at the grey-iron foundry.

It was soon discovered that the Northrop tractor fitted with the basic 67 bhp 5000 engine. was somewhat underpowered. To overcome this, the company developed a turbocharged version, which was offered in addition to the naturally aspirated model, priced at £2,495. Northrop used a CAV turbocharger which boosted engine power by 30 per cent to 85 bhp. The company encountered some initial oil consumption problems with the turbocharged tractor, but these were solved by changing to a higher quality oil. Experience gained with the turbocharger led Northrop to offer a conversion kit for turbocharging the Ford 5000 engine. The kit consisted of a CAV turbocharger, new manifold, air-breather and uprated injectors. The kit was relatively expensive, and not many were sold. A two-wheel drive 5000 on ballasted dual wheels, fitted with the Northrop turbocharger kit, underwent trials with Ford Tractor Operations and probably contributed to the introduction of their 7000 model.

The turbocharged Northrop was a good ploughing tractor with an impressive workrate but, although many more were sold than the naturally aspirated version, overall

RIGHT: Turbocharged Northrop 5004 undergoing trials with a Lemken four-furrow reversible plough.

BELOW: Turbocharged Northrop 5004 tractor. This tractor is now in preservation.

sales were still disappointing. The concept of turbocharging tractors was still in its infancy and many farmers had great doubts about its use. Some problems had also arisen when the tractor was used for pto work. The power take-off was still only designed to take 67 bhp, and when over 80 hp was put through it, breakages occurred. The Northrop had a good 18 inch ground clearance under the axles, but many farmers were put off by the overall height of the machine. It was a very manoeuvrable tractor, and its small 30 ft turning circle was much better than the equivalent County's, but in the end the County was a cheaper tractor. Only around 100 to 150 Northrop 5004 tractors were built. The expected export sales never materialised,

RIGHT: Industrial version of the turbocharged Northrop tractor.

LEFT:
Northrop turbocharging kit for the Ford 5000 engine, complete with CAV turbo, new manifold, air-breather and injectors. Note the oil-feed to the turbo.

and 95 per cent of the tractors were sold in the UK. A military rough-terrain forklift based on the tractor underwent trials with the Army, but the company received no orders for it.

In an attempt to combat falling sales, Northrop exhibited a six-cylinder version of the tractor at the 1966 Smithfield Show, designated the 5004/6. Only two prototype 5004/6 tractors were built, both fitted with Ford 2704E 112 hp engines. Other improvements included an uprated clutch and strengthened pto and gear-train. The tractor was also fitted with a semi-chassis to support the engine. The model exhibited at Earls Court also had the benefit of a Duple cab.

The production model six-cylinder Northrop 5006 was introduced at the Royal Show in July 1967, priced at £2,900. The production model differed in appearance from the prototype as it was fitted with distinctive tin work of Northrop's own fabrication, which was cheaper than the previously used Ford tin work, and could accommodate a bigger capacity fuel tank, larger radiator, and allow for greater ballast at the front. The production model 5006 was also fitted with the less powerful (90 hp) 2703E engine. The first 5006 was sold to Windsor Estates, and, unfortunately, only

LEFT:
Northrop Tractor Ltd.'s Ford D Series lorry leaves Ware loaded with two turbocharged 5004 tractors.

TOP:
One of the two prototype six-cylinder Northrop 5004/6 tractors built is seen at work in East Anglia.

ABOVE:
Another view of the prototype Northrop 5004/6 tractor. The cab was made by Duple.

three more were built before production of the Northrop tractor stopped.

By this time, British Northrop was in financial difficulties. Its loom sales relied heavily on its third-world export market which had been badly affected by both the Arab-Israeli conflict and the Nigerian civil war, and the company was facing heavy losses. The foundry closed in August 1967, and Northrop tractor production ceased the following month.

Chaseside was still a viable proposition, and was the 'jewel in the crown' in the British Northrop group. The Super-Loadmaster range had been redesigned and re-styled in 1967, and over twenty shovels a week were coming off the production lines. Joseph Cyril Bamford was looking to acquire a range of loading shovels to increase his range of JCB machines, and was able to come up with an offer for Chaseside which cash-starved British Northrop readily accepted. So, in April 1968, JCB bought Chaseside for £900,000, and production of the Super-Loadmaster shovels moved to Rochester, where they were integrated into the JCB range, remaining in production until being replaced by articulated loaders in 1972. At the time, Mr. Bamford had no interest in the agricultural sector, and did not wish to continue with the Northrop tractor production. The three remaining unsold 5006 tractors went to JCB Farms where, ironically, they worked into healthy old age without any serious problems. Many of the former Chaseside employees were

LEFT:
Production model Northrop 5006 tractor with demonstrator, Les Carter, at the wheel.

taken on by JCB, including designer Gilbert McIntosh and Les Carter, who is still with JCB as wheeled loader technical service manager.

JCB's involvement with Ford tractors is worth a brief mention. Joseph Cyril Bamford set up as an agricultural engineer in 1945, and one of his early products was the hydraulic Major-Loader, designed to fit the Fordson E27N Major. Many successive machines, including the Load-Over and Load-All

LEFT:
JCB's hydraulic Major-Loader on trial with the Ford Motor Company, fitted to the prototype E1AD Fordson Major tractor.

loaders, were either based on, or would fit, Fordson tractors.

Bamford's first backhoe-loader, the JCB Excavator MkI, built in 1953, was also based on the Fordson Diesel Major tractor, as were the subsequent Hydra-Digger-Loadall and further JCB backhoe-loaders, until the company experienced difficulty obtaining Ford skids in 1965 and moved over to using Nuffield and Leyland units.

Today, JCB is back in the agricultural market with its Fastrac range of high-speed tractors, but now using Perkins and Cummins power.

ABOVE: One of the two prototype 5004/6 tractors awaits a new owner in a dealers yard.

RIGHT: The Northrop 5006 tractor at a ploughing demonstration.

ABOVE:
JCB Load-All based on a Fordson Super Major.

LEFT:
JCB version of the Chaseside Super-Loadmaster SL 1750 shovel after its production was transferred to Rocester.

Chapter 5

Muir-Hill

Four Muir-Hill 101 tractors working with Mather and Platt pea viners.

2
2
Muir-Hill
LBE 623F

To many, the name Muir-Hill is synonymous with the dumper trucks which were a common sight on building and construction sites through the 1950s and 1960s. Although dumpers were an important part of the company's business, a large variety of machines bore the Muir-Hill name, culminating in a range of high-horsepower four-wheel drive tractors in the 1980s.

RIGHT: One of Muir-Hill's earliest products - the Model A locomotive, powered by the 24 hp Ford Model B car engine.

RIGHT: The Muir-Hill Model L narrow gauge locomotive, dating from about 1930. Powered by the Fordson tractor engine, it was capable of hauling 102 tons.

RIGHT: The Muir-Hill Light Shunting Locomotive for work on standard gauge track, fitted with the Fordson tractor engine supplied to run on petrol or paraffin.

Muir-Hill has had a chequered history, especially in the later years as the name passed from company to company. The original company to bear the name, which was set up in the early 1920s to manufacture chemical and foundry plant, was Muir-Hill Service Equipment Limited. The founders, Mr Hill and Mr Muir, working from premises in Old Trafford, Manchester, soon became involved in manufacturing Ford-powered narrow-gauge railway locomotives, and shunting tractors based on the Fordson. By the mid 1920s the company was importing attachments for the Fordson tractor from the USA, including the Iron Mule, a tipping bucket on a sub-frame, which set in motion ideas for the first Muir-Hill dumper.

At this time, Walter Hill, a member of the family of one of the founders, was working for Ford in Detroit, and was probably instrumental in arranging the importation of the American attachments. On returning to Britain, Walter Hill was employed by Rushton Tractors Limited, eventually becoming export sales manager. Thus Muir-Hill were persuaded to use the Rushton as a basis for their machines as often as the Fordson, especially as there was much similarity between the two makes. By way of circumstance, after the demise of Rushton, Walter Hill became a director of Roadless Traction before being installed as managing director of Bristol Tractors.

In late 1926, the company merged with another firm to become Muir-Hill (Engineers) Limited. The first Muir-Hill dumper, based on the Fordson Model F, was introduced in 1927, and the company further consolidated its interests in earth-moving vehicles by building a loading shovel, based on an

LEFT:
The first Muir-Hill dumper, built in 1927, based on the Fordson Model F tractor.

BELOW:
An early Muir-Hill rope-operated loading shovel. Based on the Fordson, it dates from about 1930.

American design, in 1929. The original loading shovel was wire-rope operated, and fitted to both Fordson and Rushton tractors. The main business, however, continued to be chemical and foundry plant, coupled with some interesting diversifications into surface excavation equipment, backfillers, cranes and winches. Muir-Hill also produced early tractor cabs, special wheel equipment and marketed Carrimore refuse trailers for use by municipal corporations.

The company was taken over in 1931 by one of its own directors, Edward Boydell. The new firm, E. Boydell and Company Limited, continued to use the Muir-Hill trade name. Boydell concentrated on building dumpers and loading shovels, and the company soon established an enviable reputation for its products in the construction industry.

The railway locomotives were slowly dropped from production, but Boydell continued to build shunting tractors based on Fordsons, often fitted with purpose-built Muir-Hill cabs. Many railway companies used the Muir-Hill shunters, converted from the Model N Fordson, and after nationalisation of the railways in 1948, British Rail became customers for E27N and Diesel Major based

Almost without exception, all the Muir-Hill machines used the Fordson skid unit. By the time war broke out in 1939, Boydell had sold over 1,400 dumpers and loading shovels.

During the Second World War, Boydell's production increased dramatically as dumpers were required in great numbers for the war effort. Aveling Barford built a number of larger dumpers for the Air Ministry, but most of the Grantham company's production capacity was taken up by the manufacture of Bren-gun carriers, shell caps and submarine engine components. So the licence to supply 2 cu yd dumpers went to Boydell, and the whole of its output was taken up by the Ministry of Supply for distribution to the armed forces and civil defence. The Old Trafford premises were extended to cope with the extra work, and by the end of the war, 14,000 Muir-Hill machines had been sold.

ABOVE: A 1934 Fordson-based Muir-Hill shunter built for the London and North Eastern Railway.

RIGHT: A 1931 Muir-Hill 2 cu yd dumper built onto an early Model N Fordson.

machines. The machines were also used in docks, mines and on the private sidings of large industrial companies, and were supplied up to 1959, the last units being based on the Power Major.

Without a doubt, the dumper was Boydell's best-selling product. The 2 cu yd dumper was introduced on pneumatic tyres in 1932, based on the Model N Fordson which had just gone into production at Dagenham. The company soon introduced larger capacity machines of 2½, 3 and even 3½ cu yd capacities, and were often fitted with twin wheels for increased flotation and traction in poor conditions.

After the war, a twist of fate was to put the Muir-Hill dumper on British farms. Faced with the prospect of the Government flooding the market with war surplus stocks of unused Muir-Hill dumpers, Boydell's only option was to buy up these machines and take them out of circulation. To preserve the sales of new machines to the construction industry, it was decided to modify these surplus dumpers and sell them for agricultural work. Although technically new and unused, each dumper was thoroughly checked and overhauled

ABOVE:
A Fordson-based 1938 Muir-Hill 3 cu yd dumper, fitted with twin wheels for increased adhesion.

LEFT:
An impressive line-up of wartime Muir-Hill dumpers, possibly at a dispersal sale.

RIGHT: Muir-Hill 2 cu yd dumper with a forward driving position. Based on the Fordson, it dates from the wartime.

RIGHT: Advertising the Powacart for agriculture in 1946.

LEFT: A Muir-Hill shunting tractor based on the Fordson E27N Major fitted with a Perkins P6 diesel engine.

LEFT: The Muir-Hill Powacart towing a set of disc harrows.

where necessary, and fitted with a drop-side body and a forward-facing driving position. Most were fitted with a governor and modified to run on paraffin.

Sold to the farmer from 1946 as the Powacart, they were described by contemporary Muir-Hill literature as 'one of the most useful machines which the farmer can buy, representing a prime mover for general haulage and towing and a handy vehicle of 2 tons capacity which will travel safely over any sort of ground and tip its load instantly where required.' Perhaps this was the first dual-purpose systems tractor. If the farmer was not familiar with the Muir-Hill name, he certainly knew the pedigree of the Fordson underneath, and the machine became a popular addition to the fleets of large farms as a speedy method of transporting the crop from field to yard.

RIGHT: A 1946 Muir-Hill 10B dumper. Based on the Fordson E27N Major, the driver's seat and integral steering wheel would rotate through 180° to face either direction.

RIGHT: A 1948 Muir-Hill 14B dumper, fitted with a Perkins P6 engine.

LEFT:
1959 Muir-Hill 10B dumper based on the Fordson Power Major.

Having disposed of the problem of the war surplus dumpers, Boydell introduced a new range of machines for the construction industry. The 10B, with a rotating driver's seat and controls, and the 10S were both 3 cu yd dumpers introduced in 1946, using the new Fordson E27N skid unit. The year 1948 saw the company's first move away from Ford power, with the 4½ cu yd 14B dumper. It had a Perkins P6 engine driving through a four-speed shuttle gearbox to a worm and wheel final drive. Loading shovel production continued with an improved SS model, based on the E27N, but still rope operated.

The company's wartime expansion meant that it now had the capacity to mass produce machines, and Boydell felt the need to create new markets for its products. Additional service depots were set up in Glasgow, Manchester and Birmingham as UK sales increased, but the real impetus was for overseas sales as the company went in for an export drive. Agents and distributors were appointed in sixty-five different countries, and by 1955, fifty per cent of the company's output was exported.

During the 1950s, Boydell started to investigate the use of hydraulics. Hydraulic tip was offered on the dumpers, now based on the new Fordson Diesel Major. The 2WL, the company's first hydraulic loader, was

BELOW:
The 1962 Muir-Hill 18B 6 cu yd dumper, powered by the Ford 590E engine, was the largest in the range at the time.

RIGHT: The Muir-Hill BD4 four-wheel drive loading shovel introduced in 1962.

BELOW: The skid unit for the BD4 loader. The torque converter and the transfer box are sandwiched between the Fordson Super Major engine and the final drive transmission.

introduced in 1954, also using the Fordson skid unit. A four-wheel drive loader, based on the skid-steer County Four-Drive, was brought out in 1958, but few were sold, while the 2WL went on to become one of the company's most popular products.

Production of the 10B continued on the Fordson Power Major, and later Super Major, skid units, featuring revised styling and the option of an enclosed cab. The largest model Muir-Hill dumper, the 6 cu yd 18B, was introduced in 1958. Powered by the six-cylinder Ford 590E engine, it had a five-speed gearbox, with an auxiliary box providing a forward and reverse shuttle.

In 1959, Edward Boydell sold out to Winget Limited, from Rochester in Kent. Established in 1908 as the Winget Concrete Machine Company, building cement mixers, the Rochester company had its own range of civil engineering and road making equipment which was complemented by the Muir-Hill products.

In 1962, the Winget Group moved Muir-Hill to a new twenty-eight acre site in Gloucester, formerly occupied by the Gloucester Railway Carriage and Wagon Company Limited. The Gloucester concern retained the name of E. Boydell & Co. Ltd., but traded under the Muir-Hill name.

Under Winget ownership, the company concentrated more on its loading shovel range, culminating in the introduction of the BD4 four-wheel drive machine. Following a year's testing of the prototype, production of this 1 cu yd shovel started in early 1962, soon transferring to the new Gloucester factory. Powered by the Super Major engine, it was the last Muir-Hill loader to be based on a

LEFT:
The B6000 four-wheel drive, rear-wheel steer loading shovel, the largest of the Muir-Hill range introduced in 1966.

LEFT:
Muir-Hill A5000 four-wheel drive, all-wheel steer loading shovel.

RIGHT: Muir-Hill B5000 loading shovel, fitted with a rear-mounted backhoe.

Fordson skid unit. The Fordson rear transmission unit was reversed to form the front axle of the machine, the drive for which came through a Brockhouse torque converter, and an epicyclic gearbox supplied by Roadless. The rear axle was provided by a Kirkstall drive-steer unit driven off a transfer box.

The Muir-Hill product line was completely transformed after David J. B. Brown joined the company in 1965, subsequently becoming managing director. Drawing on experience gained with drive-steer axles during his spell

RIGHT: The Muir-Hill 101 tractor introduced in 1966. The cage-wheels were built by Bettinsons of Holbeach.

LEFT: A close-up of the Muir-Hill 101 front drive-steer axle.

at Northrop, he motivated the development of seven totally new loading shovels which, incredibly, were in production by 1966. The range included the two-wheel drive F4000 and F5000, and the four-wheel drive B4000, B5000 and B6000 models. The A4000 and A5000 machines had the advantage of two drive-steer axles giving them all-wheel steer as well as four-wheel drive, and the benefit of a very tight turning circle. The model number referred to the machine's payload - for example, the F4000 would lift 4,000 lb.

All the new machines were fitted with a long, single, centre-mounted lifting beam, forged from heavy gauge steel, and an off-set cab giving maximum visibility. The only Ford content in the loading shovels was the 2711E engine in the 4000 series shovels. Allison Torqmatic Powershift transmissions were fitted and the larger machines had Leyland and, later, Perkins power units.

David Brown also initiated the design of a new four-wheel drive tractor, the Muir-Hill 101, introduced in November 1966. The

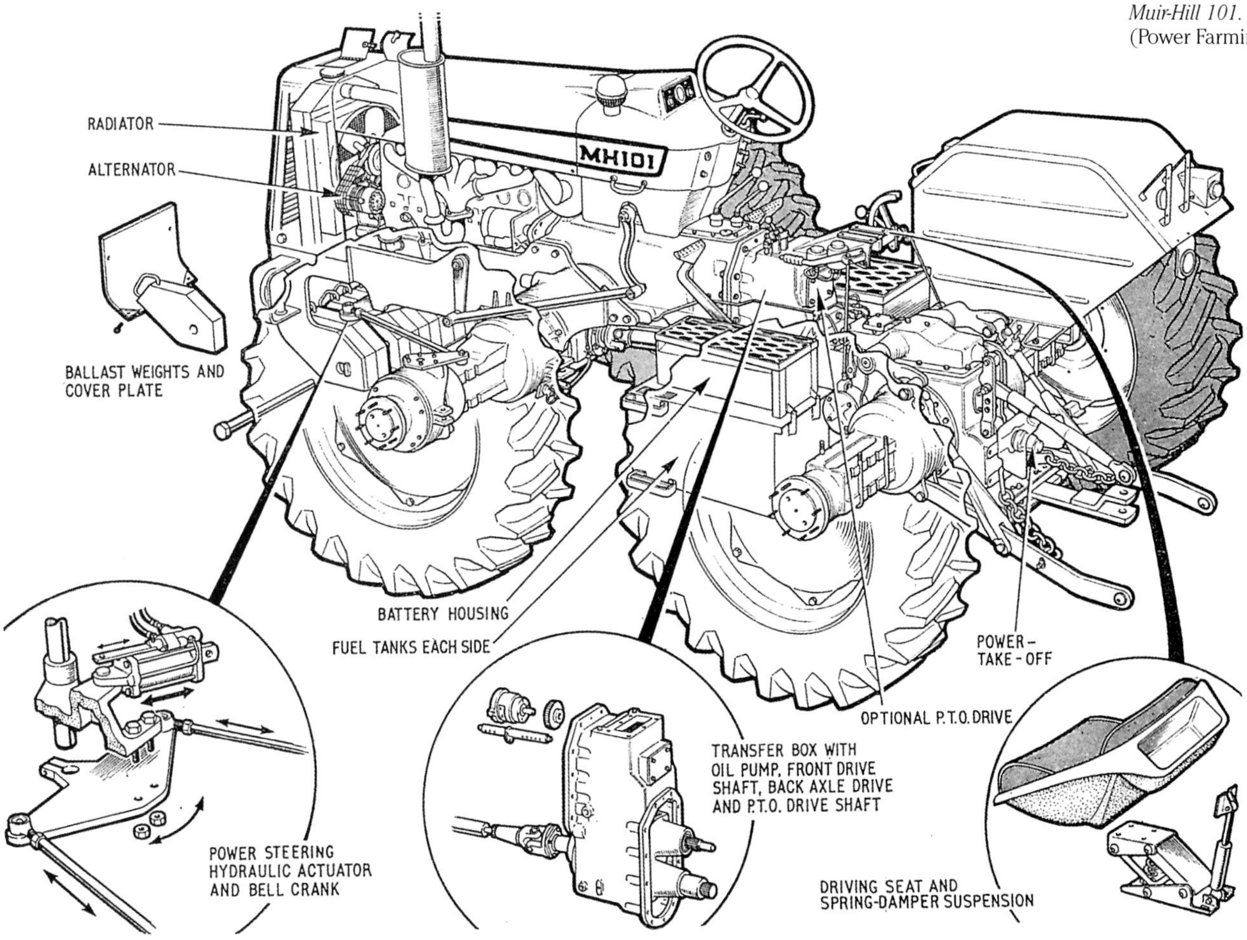

BELOW: An exploded view of the Muir-Hill 101. (Power Farming *copyright)*

RIGHT:
A pair of Muir-Hill 101 tractors. The tractor nearest the camera has a factory fitted cab, while the frame on the machine on the right appears home-made.

RIGHT:
A Muir-Hill 101 on dual wheels cultivating with a ripper plough in Kent.

tractor was conceived as a universal machine, with both agricultural and industrial uses, that would also appeal to overseas timber and sugar cane growing areas.

Built along similar lines to the Northrop, the 101 was fitted with a Ford 5000 gearbox, with a transfer box splitting the drive equally to the front and rear axles. The transfer box was lubricated by its own oil pump, and also drove the power take-off. The rear transmission and hydraulics were derived from the Ford 5000, but the front drive-steer axle was taken from the smaller Muir-Hill loading shovels. Power was provided by a six-cylinder Ford 2704E engine rated at 108 bhp. The price was set at £2,850 for the agricultural tractor, with the

ABOVE:
A Muir-Hill 101 fitted with a sprayer attachment built by a Yorkshire farmer. This set-up was capable of covering an acre in 1 minute 40 seconds.

LEFT:
Preparing a seedbed for potatoes with a Muir-Hill 101 and a Howard Rotavator.

RIGHT: Two Muir-Hill 101 industrial tractors fitted with Boughton winches.

BELOW: The Muir-Hill low-speed tractor is fitted with a reduction box to enable it to work this Norwegian-built Radahl trencher on tile drainage.

LEFT:
Laying plastic drainage pipe with a Muir-Hill 101 and Radahl trencher.

LEFt:
Muir-Hill XL200 mobile crane with its own cruciform base.

industrial version priced at £2,750.

With excellent weight distribution and plenty of power, the Muir-Hill tractor was popular with farmers and contractors alike, and soon proved its worth out on the big farms. Two were bought by a farm in the East Riding of Yorkshire as ploughing tractors. Matched to Doe eight-furrow ploughs, each tractor would average a workrate of 2.5 acres an hour, or cover 100 acres a day with an 18 ft springtine cultivator on seed-bed preparation. Its impressive turning circle of a little over 16 ft meant that the 101 was manoeuvrable enough to be used to pull potato harvesters and pea viners when conditions demanded extra traction.

The industrial version found many applications, including site haulage and timber handling, as well as being fitted with winches, cranes and dozer blades. A low-speed version of the 101 was brought out in 1967, designed for trenching, sub soiling and land clearance. It had a 24:1 reduction gearbox and sixteen gear ratios providing speeds from 0.075 to 19 mph. A few 101s were turbocharged by Colchester Tractors.

FJB 130K
M·H110

Muir-Hill
MH
161

LEFT:
Muir-Hill 161 with a 25TD dumper trailer.

FACING PAGE TOP:
The uprated Muir-Hill 101 tractor, introduced in 1971.

FACING PAGE BOTTOM:
The massive V8 Perkins powered Muir-Hill 161 tractor.

The Muir-Hill title changed hands again in 1968, when the holding company, Winget Gloucester Limited, was bought by Babcock and Wilcox Limited, specialists in heavy mechanical engineering with a reputation for enterprising research and technological development. Under Babcock's ownership, E. Boydell and Company Limited was renamed Muir-Hill Limited, in turn becoming part of Babcock UK Investments Limited. The Muir-Hill range was revised to include the loading shovels, the 101 tractor, two dumpers and a new self-propelled hydraulic crane which had its own cruciform base.

Two new tractor models were brought out in 1969. The 110 was a Perkins-powered version of the 101, fitted with the six-cylinder 6.354 engine. It was aimed more at the export markets that preferred Perkins engines to Ford power. The choice of engines in certain overseas territories was often dictated by what servicing arrangements were established, and some countries only had facilities for Perkins engines. Perkins-powered versions of Muir-Hill tractors were particularly popular in the sugar cane growing areas of Barbados and Guyana. Bookers Sugar Estates in Guyana ran a fleet of thirty Muir-Hill tractors, and a subsidiary, Bookers Stores Limited, became distributors for the machines.

BELOW:
Muir-Hill 111, powered by the Perkins 6.354 engine, seen on a show stand in 1973.

ABOVE:
Muir-Hill 161, powered by the Perkins V8.510 engine, with a Dowdeswell DP6 plough.

RIGHT:
Muir-Hill 101 tractor with a 20TD dumper trailer. Payload for the integrated unit was 20 tons.

The 161, the other tractor introduced in 1969, was also Perkins powered. It was a much larger machine and was fitted with the V8.510 engine developing 163 bhp. This massive beast, with its twin exhaust pipes, was fitted with a ten-speed Fuller close ratio gearbox. It had a more spacious detachable cab and hydrostatic steering. The most powerful tractor available in the UK at the time, it was aimed at the big farmer or contractor and the heavy industrial user.

David J. B. Brown introduced a new concept in dumpers in 1969. The integrated tractor/trailer dumpers were available in 13 and 20 ton sizes hauled by the 101 or 110 tractors, and a 25 ton model available with the 161. The tandem axle dumpers were attached to the tractor by a Hydra-hitch which allowed quick coupling and positive weight transference from the dumper to both tractor driving axles. The complete unit was braked on all eight wheels.

The Babcock group was reorganised in 1971, and Muir-Hill became part of the

LEFT:
David J. B. Brown at the wheel of a Muir-Hill 161 with a 25TD integrated dump trailer.

Construction Equipment Division. By this time, over 37,000 Muir-Hill machines had been sold and the range now consisted of nine loading shovels, two dumpers, two cranes and the tractors.

In 1971, Muir-Hill uprated the 101 to 120 hp by fitting the Ford 2715E engine, and the power-assisted steering was replaced by full hydrostatic. The following year, the Series II Muir-Hill tractors were introduced. The 101 was replaced by the 121, still using the 2715E engine and Ford gearbox, but now with several improvements, including a larger integral fuel tank and a better cab. The new flat-floor cab was rubber-mounted to cut down noise and vibration, and had sliding doors for easier access. The 110 was replaced by the 111 incorporating the same design improvements as the 121 but powered by the 110 bhp Perkins 6.354 engine.

The 161 stayed in production until 1975, when it was replaced by the 171, fitted with the Perkins V8.540 engine, developing a useful 170 bhp. It, too, featured similar cab

LEFT:
In 1971, the Muir-Hill 101 was fitted with the 2715E engine and uprated to 120 bhp.

RIGHT: The Earthmill , the Muir-Hill MH-1250 Rotary Excavator, capable of digging 1,000 cu yd of soil per hour.

RIGHT: Muir-Hill 171 on demonstration with a Dowdeswell DP7 five-furrow plough.

LEFT:
Muir-Hill 141 tractor, available with either turbocharged Ford or Perkins engine.

BELOW:
Muir-Hill's stand at the 1978 Royal Show. On display are agricultural and industrial 141 and 171 tractors, a B5000 loading shovel, and a rough terrain forklift. A 121 tractor can be seen in the background.

RIGHT:
One of two Muir-Hill 101 tractors fitted with aqueous ammonia injection equipment in Lincolnshire.

improvements in line with the other two models in the range, and the distinctive twin exhaust pipes were replaced by a single stack. As an interesting price comparison, less than ten years after the first Muir-Hill tractor was introduced, this top model cost £16,400.

Further rationalisation took place at Babcock in 1978. Babcock Construction Limited was formed in January, out of the old holding company, Winget Gloucester Limited, to embrace all the businesses which were previously part of the Construction Equipment Division. The parent company, Babcock International plc, was now a multi-national engineering organisation with an annual turnover exceeding £800 million, and employing 40,000 people worldwide. The new subsidiary company drew together several famous names in the construction field, including Muir-Hill, Winget, Blaw-Knox and Allatt, with a collective turnover of £70 million a year, half of which was in exports.

Each company that was part of Babcock Construction Limited pooled its resources in engineering, design and marketing to provide a complementary range of equipment. Blaw-Knox, dating as a company from 1921, had been manufacturers of road pavers since 1956. The company also supplied sweepers and distributed Dynapac rollers. BPR tower cranes and Marion Dresser draglines and mining shovels were marketed by Babcock Crane & Mining Division.

Apart from building cement mixers, Winget now concentrated on dumpers, vibratory rolls and industrial rough terrain forklifts.

FACING PAGE TOP:
A Muir-Hill 171 tractor ready for work with a three-leg subsoiler.

FACING PAGE BOTTOM:
A Series III Muir-Hill 171 with the improved 'Spacecab'.

RIGHT:
Three Muir-Hill 121 tractors pea vining in Lincolnshire with Mather and Platt machines

The agricultural version of the forklift was made by Muir-Hill, illustrating the strong in-roads it had made into the farming sector. The company was still manufacturing its range of nine loading shovels, and a unique earthmoving machine, the Earthmill, which was capable of digging up to 1,000 cu yd of soil per hour. But the four-wheel drive tractors had increasingly become a very important part of Muir-Hill's business, both in agricultural and industrial sales.

The Series III range came out in 1978, fitted with the new Spacecab, which was slightly larger with improved noise insulation and kept dust-free by an air filter unit mounted on the roof. Power was also slightly increased - the 121 was now rated at 132 bhp, the Perkins-powered 111 at 116 bhp, and the V8 in the 171 tractor now had the benefit of 177 bhp.

A fourth model was added to the range in the form of the 141, which was available with the choice of either Ford or Perkins engine. Both the six-cylinder engines offered for the 141 were turbocharged - the Ford 2704ET developed 143 bhp, while the Perkins T6.354.4 was rated slightly less at 132 bhp. The tractor was priced the same with either engine option at £25,413.

Dual Power transmissions had come in at the end of the production run for the Series II, and were now optional on the 111, 121, and 141 which still used Ford gearboxes. The 171 agricultural tractor had an Eaton-Fuller T4510 gearbox, and an Allison MT640 automatic transmission was used for the industrial model. All the tractors were available as industrial versions without hydraulic linkage, and could have optional heavy-duty towing hitches and air brakes.

Tractor production was to sadly cease in October 1982, when Babcock International decided to revise its areas of operation and sold Babcock Construction Equipment Limited to the West German firm, Wibau AG. Wibau was part of the German-owned IBH group, which unfortunately ran into difficulties and failed in November 1983. Several

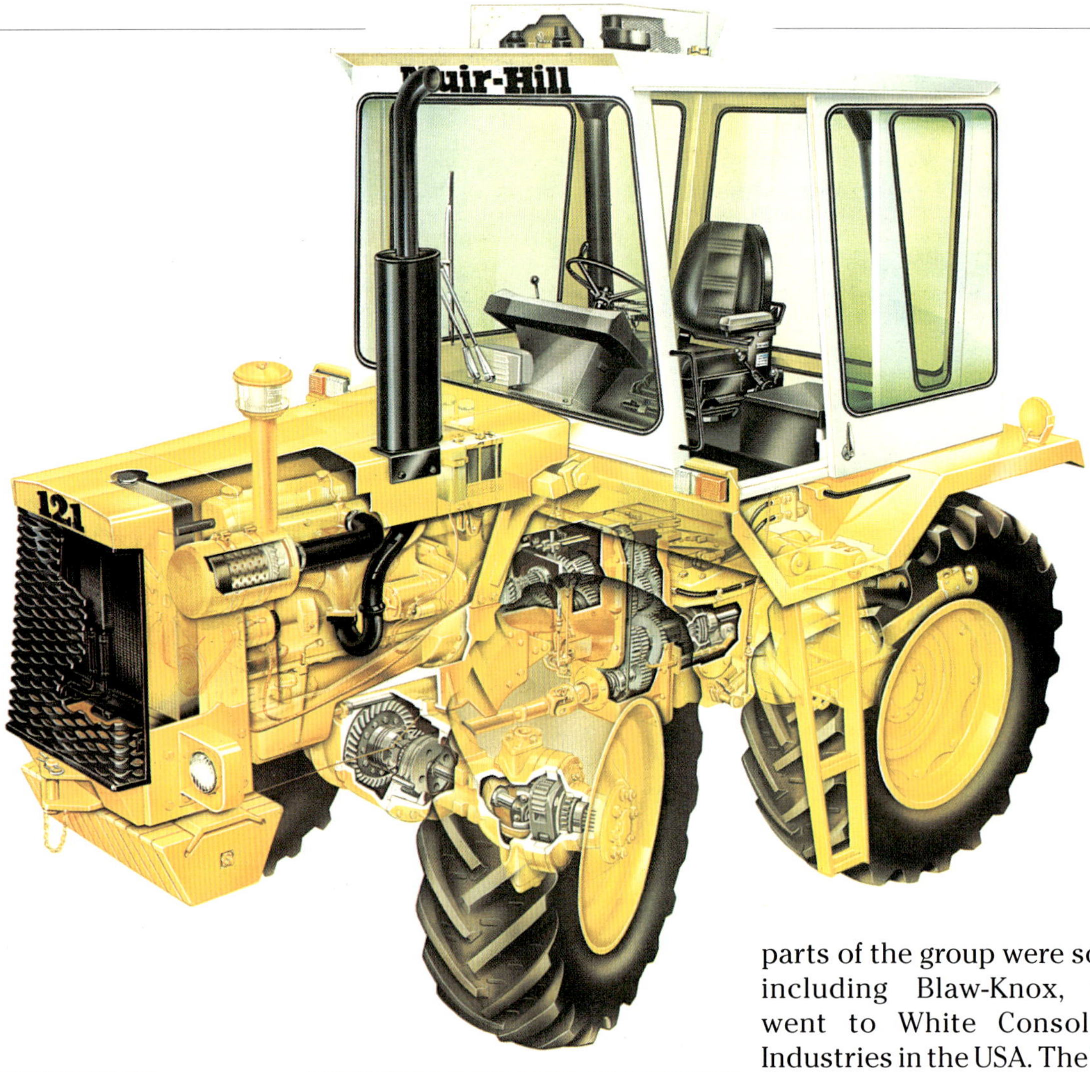

RIGHT: An exploded view of the Muir-Hill 121 Series III tractor

BELOW: Myth-Holm 131 tractor built for drainage work in Norway, fitted with a 24:1 County reduction box.

parts of the group were sold off, including Blaw-Knox, which went to White Consolidated Industries in the USA. The Wibau Winget subsidiary, which contained Muir-Hill and Winget, was run by the receivers until it was sold, in June 1984, to Sandersons Forklifts of Croft, near Skegness in Lincolnshire.

Sanderson (Forklifts) Limited was founded in 1967, by Roy Sanderson to manufacture tractor-mounted forklift attachments. The company had evolved from a blacksmith and wheelwright's business set up by the founder's grandfather, Harry, in 1912. Sanderson's first rough terrain forklift, the SB 5000 built in 1971, used a Ford skid unit. Most of the company's subsequent machines, including the Teleporter telescopic handler, were Ford based, and by the 1980s, Sanderson had an extensive range of forklifts which was further complemented by

LEFT:
The Myth-Holm 131 agricultural tractor, powered by the Ford 2725 engine developing 130 hp.

BELOW:
Lloyd Loaders' prototype A5500 all-wheel steer loading shovel brought out in 1989.

the addition of Muir-Hill and Winget products.

Sanderson completed the building of three Muir-Hill 171 tractors that still remained unfinished at Gloucester. The tractors were transported to Croft for painting, and then sold, minus cabs, to British Steel. The three tractors went to work in the Ravenscraig steel works in Glasgow, where they were fitted with cabs built by British Steel, and used for haulage.

Agricultural engineering was now an uncertain world, and in November 1990, Sandersons, too, went into liquidation. Winget was sold to Seddons (Plant & Engineering) Limited, from Bolton in

LEFT:
Lloyd Loaders' Mytholm Works, near Halifax, now proudly bears the Muir-Hill name.

RIGHT:
A Muir-Hill 141 draws a Ransomes TSR 109 five-furrow reversible plough.

Lancashire, and Sanderson and Muir-Hill were bought by the long-established Lincolnshire engineering company, Aveling-Barford of Grantham, who were involved in the manufacture of road-rollers, dumpers, loading shovels and graders. Production of the Sanderson machines was transferred to Grantham, and still continues, with a new pivot-steer farm handler released at the end of 1994.

In September 1991, the Muir-Hill business was purchased from Aveling Barford by the Yorkshire company, Lloyd Loaders (M.H.) Limited. Set up in Bradford in 1960, the company was originally involved in plant hire, demolition, excavation and quarry work. Moving to the present Mytholm Works, on a 21 acre site in Hipperholme, near Halifax, in 1966, Lloyd Loaders became specialists in rebuilding and reconditioning Chaseside loading shovels, later turning to Muir-Hill machines after Chaseside was swallowed up by JCB.

Over the years the company became more involved with Muir-Hill products, and carried out several conversions on the loading shovels, culminating in the production of their own machine in 1985. Known as the Myth-Holm B6500, this four-wheel drive, rear-wheel steer shovel was based on the Muir-Hill design with several modifications. It had a Perkins engine, Allison transmission and Muir-Hill axles.

The following year, Lloyd Loaders were approached by a Norwegian company to build a tractor of similar design to the Muir-Hill for drainage work. The prototype, designated the Myth-Holm 131, was completed in February. Fitted with a Ford 2725 engine, gearbox and Dual Power transmission, it also had a County 24:1 reduction box, giving a total of thirty-two forward and eight reverse gears. No hydraulic linkage was fitted as none was required for the drainage work. After evaluation trials, the tractor was shipped out to Norway in September. Five more were built

LEFT:
The Myth-Holm 131 agricultural tractor on demonstration in October 1989, ploughing in Suffolk.

for the same company between 1987 and 1990.

At the 1989 Great Yorkshire Show, Lloyd Loaders launched an agricultural version of the 131 tractor priced at £38,500. Developing 130 hp from the industrial six-cylinder Ford 2725 engine, it had a Dual Power transmission giving sixteen forward speeds, and an air-conditioned cab. The company was not aiming for large-scale production with the tractor, but expected to build each machine to customers' individual requirements.

In 1989, the company built a new prototype four-wheel steer 4 x 4 loading shovel - the A5500, with a Perkins engine and Allison transmission – and it continues to develop and improve its machines, which now carry the Muir-Hill name. Today, Lloyd Loaders builds new tractors and loading shovels to order, and sells and reconditions used machines. The managing director, David Lloyd, sums up the philosophy of the company in his comment, 'We don't sell our machines, we allow people to buy them.'

CHAPTER 6

Matbro & Bray

John Suckling's Bray Centaur at work in 1963.

In November 1973, Bray Construction Limited was purchased by Matbro, and the two respected firms, long established in the manufacture of wheeled loaders and materials handling equipment, became part of the same group. Each company was pioneering in its own respective field: Bray was credited with producing the first hydraulic industrial loader in the UK in 1946, and Matbro designed and patented an innovative

RIGHT: 1949 Bray Hydraloader.

RIGHT: Fordson E27N Major fitted with the 1949 Bray Pupdozer

LEFT: 1949 prototype Braydozer based on a County CFT crawler.

pivot-steer system for wheeled loaders in the late 1950s.

Bray was the older of the two companies, founded in 1911 by W. E. Bray as machine tool manufacturers, initially building components for motor torpedo boats used in the First World War. Gradually the company started to specialise in equipment for the earthmoving industry.

In 1934, Bray produced its first hydraulically operated angledozer. Eighteen of these were supplied to the Ministry of Defence, and eventually saw action in France and the Middle East during the Second World War. Throughout the war years, Bray was involved in making parts for tanks and military vehicles and reverted to construction machine manufacture after the war ended. In 1946, the company built its first loading shovel fitted to a David Brown tractor.

Bray started pioneering development work on an industrial hydraulic loader during the late 1940s. Known as the Hydraloader and based on the Fordson E27N Major Skid Unit, it was released on the market in 1949. It had single acting lift rams, parallel linkage and a cable release for the 7 to 8 cu yd bucket. This was Bray's first use of the Fordson tractor, and during the same year, it went on to build a small pup-dozer attachment for the E27N Major, fitted on an A-frame and operated by rods connected to the rear lift.

Also in 1949, Bray fitted its heavy-duty hydraulic angledozer, known as the Braydozer, on to a County CFT crawler. This proved to be a very successful machine, and led to a long association between Bray and County.

As orders for the new machines came in,

BELOW: This photograph clearly shows the rear hub extension forming the pivot-point for the angledozer. The hydraulic oil reservoir can be seen under the driver's seat.

cally, claimed as another first by the company. The new Hydraloader was designated the HL21. The HL27, a front-wheel drive version of the machine, with rear wheel steer, was also built but did not catch on.

Throughout the 1950s, Bray made further improvements to the Hydraloader including new linkage, pump changes and the introduction of high-lift versions. A range of attachments was also offered for the loader, such as a snow plough, bulldozer, angledozer, fork-lift, special buckets, crane, rock fork and barrow hoist. In 1954 the Dualoader was introduced, which allowed the bucket to pass over the cab and discharge to the rear so as to speed up the loading cycle.

ABOVE: Perkins-powered County CFT fitted with the Braydozer attachment.

RIGHT: The 1952 Bray HL21 Hydraloader, based on the new model Fordson Major, now fitted with 'crowd' rams to operate the bucket

the company was forced to expand in 1950 to a new site at Faggs Road, Feltham, Middlesex. This site, close to London's Heathrow Airport and eventually covering some ten acres, was to be home to all Bray's manufacturing and assembly plants for over twenty years.

With the introduction of the new model Fordson Major in 1952, Bray took the opportunity to improve the Hydraloader by fitting 'crowd' rams to operate the bucket hydrauli-

Bray also built the Hydraloader on Roadless and County versions of the Fordson. The Bray Half-track Hydraloader came out in 1955, based on a Fordson Major, fitted with Roadless DG half-track equipment, in an attempt to provide a machine with increased adhesion for working under adverse conditions. Conversely, in 1959, Roadless introduced its own loading shovel which consisted of the Bray Hydraloader built around a

LEFT: The HL27 reverse-drive forward control version of the Bray Hydraloader.

BELOW: A 1957 Mark II version of the Bray Dualoader shows its ability to load over the cab and discharge to the rear.

Roadless four-wheel drive Fordson Major tractor. A Sherman backhoe excavator, an American-built attachment marketed by Bray, was offered by Roadless as an optional extra, and could be fitted to the rear of the loading shovel.

The Sherman backhoe was sold by Bray as the Longford Digger, and was usually fitted to Fordson tractors, although it would go on a variety of other machines. The American manufacturers of the backhoe, Sherman Products Inc. of Royal Oak, Michigan, were bought by the Ford Motor Company in May 1960, and their products formed the basis of Ford's industrial machines in the USA.

The Braydozer attachment was fitted to most County models in the late 1950s and early 1960s, including the crawlers, Four Drive and Super-4 tractors. The Hydraloader was fitted to the industrial County Mk III, Mk IV and CD 50 crawlers, and sold as the Brayloader BL25.

In 1955, Bray Construction Equipment was bought by Sheepbridge Engineering Limited. Originally, Bray's major output had been angledozers and bulldozers fitted to a variety

RIGHT:
A Bray Dualoader loads into an early Aveling Barford shuttle dumper.

BELOW:
The Bray Half-track Hydraloader, based on a Fordson Major fitted with Roadless DG tracks, on demonstration in a clay pit near Stoke-on-Trent, 1955.

of different units, including Fowler, Track Marshall and Fiat crawlers, as well as the Ford based machines, but now loading shovels were accounting for nearly half of the company's production. With the strength of the new engineering group behind them and taking advantage of the facilities they offered, Bray decided to introduce a re-designed range of loaders for 1956.

Still based on the Fordson Major skid-unit, they featured re-styled, enclosed bodywork with the advantage of a fibreglass weather cab. Power steering became standard, and a maroon and cream colour

LEFT:
The Roadless four-wheel drive loading shovel, fitted with a Bray loader, photographed at Hounslow.

scheme, similar to British Railway's 'plum and custard', was adopted. The Hydraloader became the BL30, the high lift version was numbered BL32 and the Dualoader was re-badged as the DL35. The County BL25 completed the range of four new loaders.

Also in 1956, Bray developed the BL430, probably the first four-wheel drive industrial loading shovel on the UK market, again based on the Fordson skid unit. The first two machines, built with four-wheel drive axles supplied by All Wheel Drive of Camberley, Surrey, were no great success and were sold

LEFT:
The Sherman backhoe, seen fitted to an American Ford tractor, was marketed for a while by Bray.

LEFT:
1954 County Mk III crawler fitted with a Braydozer attachment.

RIGHT:
County Four Drive skid-steer tractor with a Bray angledozer that was offered as original equipment with the machine.

BELOW:
The Bray BL25 loading shovel, based on the County crawler, was introduced in 1956.

LEFT:
One of the new range of shovels introduced in 1956, the Bray BL30.

LEFT:
The 1956 Bray BL35 Dualoader.

to Germany. BMC axles were tried next, but by 1958 Bray was developing its own four-wheel drive axles, using BMC differentials and manufacturing its own axle cases and hub reductions. A similar version, known as the BL 432T, saw the company's first use of a torque converter transmission, which was supplied by Brockhouse.

The four-wheel drive BL 440 loading shovel, built in 1957, was a purpose-built machine, and Bray's first loader not to be based on a Fordson Major skid unit, although the Ford engine and gearbox was still used. With equal-size wheels, rear-wheel steer, a forward driving position, and a distinctive fibreglass cockpit and bodywork, it looked like no other Bray machine that had gone before it. The successive BL 450, BL 455 and BL 460 machines were offered with the option of torque converter transmissions, and a choice of Ford, Perkins or Leyland engines, as the company began to move away from the sole use of Ford power.

By 1958, Bray was manufacturing its own hub reduction units, power-shift gearboxes and fluid

RIGHT: The Bray BL35 Dualoader loads a 1944 Bedford OWST truck.

RIGHT: One of two Bray BL 430 four-wheel drive loading shovels fitted with All Wheel Drive front axles, built in 1956.

transmissions, and the following year, introduced external disc brakes, built by Sheepbridge Engineering, on to its machines. In 1961, the company was rewarded by winning a contract worth £1 million with the Ministry of Defence to supply machines and spares, including the BL 455 loading shovel, the BL 455 MW (an amphibious version) and Sherman backhoes.

Ford components were never phased out completely, and when a new range of loading shovels was introduced in 1965, the three smallest models, the BL 530, BL 534 and BL 536, used the Ford 5000 engine. In fact, Bray loaders used Ford engines into the 1980s, alongside Perkins, Leyland, Cummins and Volvo power plants.

Bray's involvement in agriculture started

LEFT:
The 1957 Bray BL 440 four-wheel drive loading shovel, fitted with Ford engine and transmission.

BELOW:
1965 Bray 536 loader fitted with the Ford 5000 engine.

ABOVE:
Bray 450 TF loading shovel on the forecourt of Ford dealers, Crimble of Staines.

RIGHT:
The original Bray Centaur today, awaiting restoration.

almost by chance when the company was approached in 1958 by John Suckling, an agricultural engineer and contractor from Steeple Bumpstead in Essex, to develop a four-wheel drive ploughing tractor with a plough mounted at either end. His idea was not the modern concept of the 'push-pull' plough, but a system whereby only one plough would be in work at once. The method of ploughing would be to drive forwards down the furrow with the rear plough in work, and then raise this, lower the front plough and reverse back up the furrow, with no need to turn on the headlands.

Suckling's inspiration for the concept, which he saw as the ideal solution to covering more acres per day when contract ploughing, came from working with steam balance ploughs. He had originally taken the idea to Blaw Knox, with the notion of developing one of their graders into the machine he required, but this was found to be unsuitable.

Bray were interested enough to lend John Suckling a BL 450 TF shovel skid unit on which to base his machine, and he built up the ploughs for it in cooperation with Ransomes, using their three-furrow TS73 plough converted to five furrows. The frame for the

ABOVE: John Suckling photographed with his Bray Centaur in 1994.

LEFT: The Ford 590E 96 hp power plant in the Bray Centaur.

ABOVE: The prototype Bray Centaur ploughing tractor at work

RIGHT: Inventor John Suckling at the wheel of the Bray Centaur.

front plough was turned upside-down, and the legs were altered to accommodate left-hand bodies taken from a reversible plough.

Originally the plough was set to pull 10 in. furrows, but this was later altered to 11 in. so it would pull more central to the tractor. A four-way hydraulic control valve, fitted to the tractor, controlled lifting and lowering of the ploughs, and also had a float position for use when a plough was in work.

The resulting prototype was used by John Suckling in development trials during 1958. The Bray machine had retained its original torque converter transmission, and this was

ABOVE:
The Bray Centaur was designed to work with a plough mounted front and rear.

LEFT:
Front view of the Bray Centaur reversing up the furrow. The front mounted five-furrow plough was fitted with left-hand bodies.

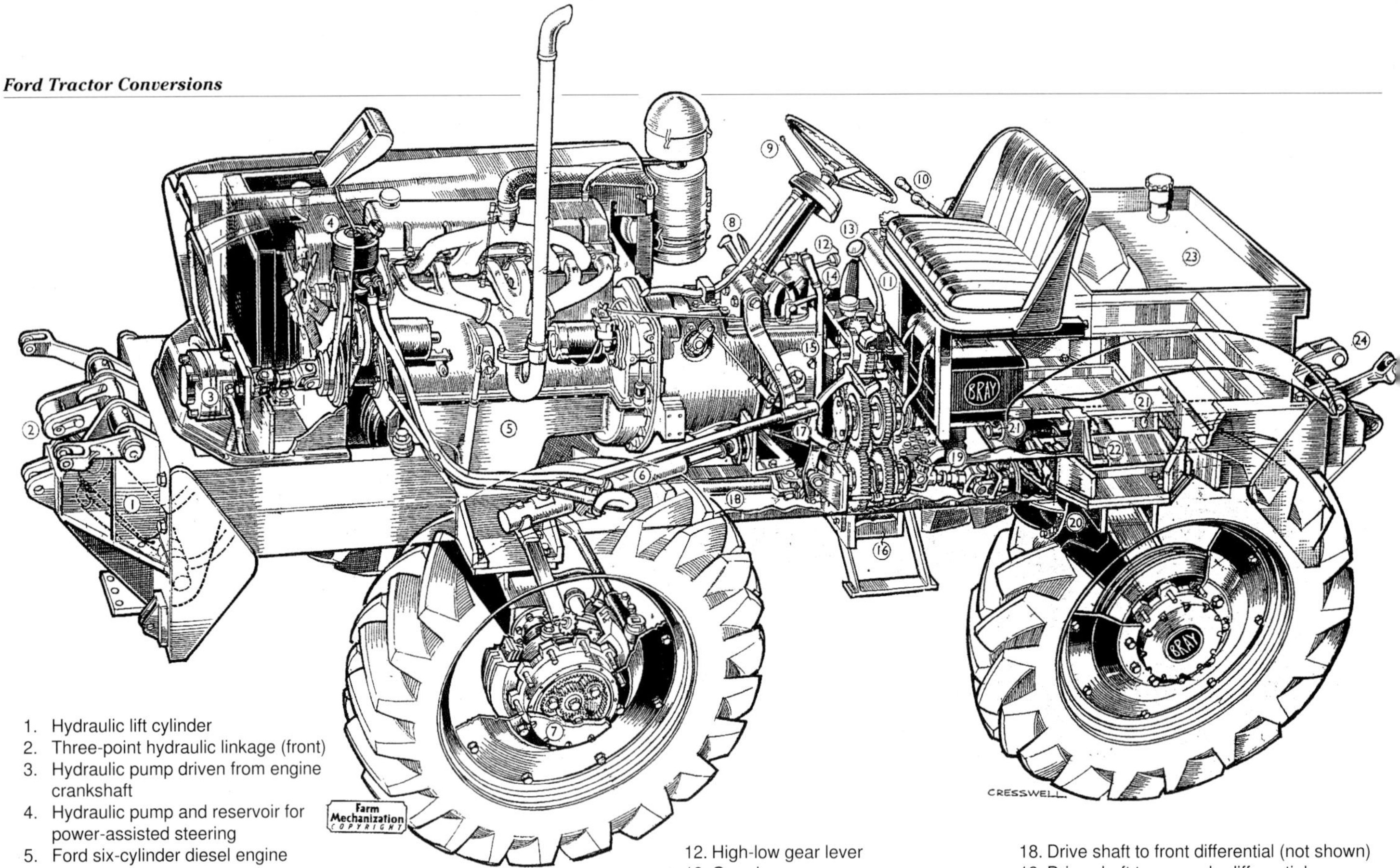

1. Hydraulic lift cylinder
2. Three-point hydraulic linkage (front)
3. Hydraulic pump driven from engine crankshaft
4. Hydraulic pump and reservoir for power-assisted steering
5. Ford six-cylinder diesel engine
6. Hydraulic steering ram
7. Epicyclic reduction gear
8. Hand Brake
9. Throttle
10. Control levers for front and rear hydraulic lifts
11. Hydraulic oil tank
12. High-low gear lever
13. Gear lever
14. Electric starter lever
15. Transfer box gear lever controlling forward and reverse speeds
16. Transfer gearbox
17. Clutch pedal
18. Drive shaft to front differential (not shown)
19. Drive shaft to rear axle differential
20. Rear differential housing
21. Oscillating axle pivot
22. Axle stop
23. Fuel tank
24. Three-point hydraulic linkage (rear)

ABOVE: Exploded view of the Bray Centaur tractor. (Farm Mechanisation *copyright*)

RIGHT: Bray Centaur with the four-furrow digger plough based on the Ransomes TS59.

LEFT:
Bray Centaur working an 8 ft seven-tine heavy duty cultivator to a depth of 12 in.

BELOW:
Bray Centaur pulling a twin-leg subsoiler to a depth of 22 in.

found to be unreliable and very heavy on fuel, and it was impossible to keep the oil cool when ploughing for long periods, even with up to ten oil-cooling radiators strung around the tractor.

On John Suckling's advice, the torque converter was taken out, and a normal Fordson Power Major manual gearbox was fitted. A forward-reverse shuttle was achieved by fitting a dog clutch to the transfer box. With this direct-drive, fuel consumption was nearly halved to about two gallons per hour, and drawbar-pull was greatly increased. A stabiliser disc was also fitted to the rear of the tractor to aid keeping up to the furrow wall when using the front plough.

At the end of 1958, the tractor was returned to Bray, who then started development work on a production model which was based on a BL 455 loading shovel, and fitted with a 96 hp Ford 590E engine. The new machine was to be called the Bray Centaur, although the designers back at the Bray works had dubbed it the Suffolk Special (John Suckling lived close to the Suffolk border, and much of the development work was done in that county).

The first tractor was delivered to John Suckling in September 1959, with the agreement that he would demonstrate the machine for Bray and build the implements for it. The tractor was priced at £2,527, plus £150 for the category II three-point linkage, front and rear. The complete ploughing machine, with two five-furrow ploughs, cost £3,267. Suckling demonstrated the tractor all over the country, including Scotland and Wales, and generated considerable interest, normally ploughing at the rate of two acres per hour.

John Suckling built a whole range of implements for the tractor, including a four-furrow digger plough based on the Ransomes TS59, a seven-tine heavy duty cultivator and a twin-leg subsoiler. A 1 ton cast iron counterbalance

RIGHT: When working with rear mounted implements, the Centaur had a 1 ton cast weight mounted on the front linkage.

weight was fitted to the front linkage when the tractor was used with rear-mounted implements. A loading shovel could also be mounted on the front of the tractor.

A delegation from the West Indies came twice to inspect the tractor for sugar cane work, believing it to be the ideal solution for replacing both the heavy crawlers used for cultivation work, and the Marshall MP6 tractors employed in road haulage, with one universal machine. They were initially interested in ordering twenty tractors, and a Centaur was sent out to the West Indies on trial with a special disc ridger, built up by John Suckling in conjunction with Ransomes.

Reports on the tractor's performance were

BELOW: John Suckling also developed a front-end loader attachment for the Centaur, with the rear cultivator acting as a ripper for site work.

LEFT:
The production model of the Bray Centaur, complete with weather cab.

LEFT:
Prototype Bray Four 10/60 tractor. The extended front tinwork may have been used to house the experimental six-cylinder engine, possibly in an attempt to emulate the County Super Six.

favourable, but a better steering lock and an increased road speed of 20 mph were required. John Suckling suggested using two drive steer axles to give crab steering, but Bray considered the modifications too expensive, and the order never went through.

By now, Bray were too heavily involved with the large Ministry of Defence contract to spend time and money on developing the Centaur further. Sales had not really materialised as the machine was expensive, and more interest was shown in an industrial

RIGHT: Six-cylinder version of the Bray Four Nuffield - possibly the prototype.

version available with an angledozer, Sherman backhoe or 10 ton logging winch.

In the end, two were sold in Scotland, two were sent to Brazil to work on coffee plantations and were paid for in coffee beans, and several went to Babcock & Wilcox as prime-movers for heavy-duty pressure vessels. The best customer was Weston Excavating Limited, a division of English China Clay, who used it as a recovery vehicle, fitted with an angledozer at the front and a winch at the rear, and continued to order units up to 1973. Trials were made with the Ministry as an army towing vehicle, but none were actually sold.

RIGHT: The production model Bray Four 10/60, based on the 60 hp Nuffield.

LEFT:
The underpan, evident in this photograph of the Bray Four 10/60, forms part of an A-frame mounting for the front axle, as well as protecting the prop-shaft and drop-box.

LEFT:
An industrial Bray Four 10/60 fitted with a Norwegian Hymas loader.

The original Centaur was kept by John Suckling for use in his contracting business, and is still owned by him today and is in full working order. Each year for eighteen years, until 1976, when safety cab laws made its use difficult, the tractor would plough 1,000 acres in ten weeks. Outside of the ploughing season, the tractor was used with a front-end loader on site work, or was fitted with a snow plough. The engine was only overhauled once, but the gearbox was regularly rebuilt, eventually being fitted with heavy-duty bearings by John Suckling, who claimed he could completely remove the engine and gearbox in half an hour. The tractor is currently under restoration.

Although the remainder of Bray's agricultural machines were not based on Ford tractor skid units, and therefore technically outside the scope of this book, it would be

RIGHT: Exploded view of the Bray Four 4/65 tractor. (Power Farming copyright)

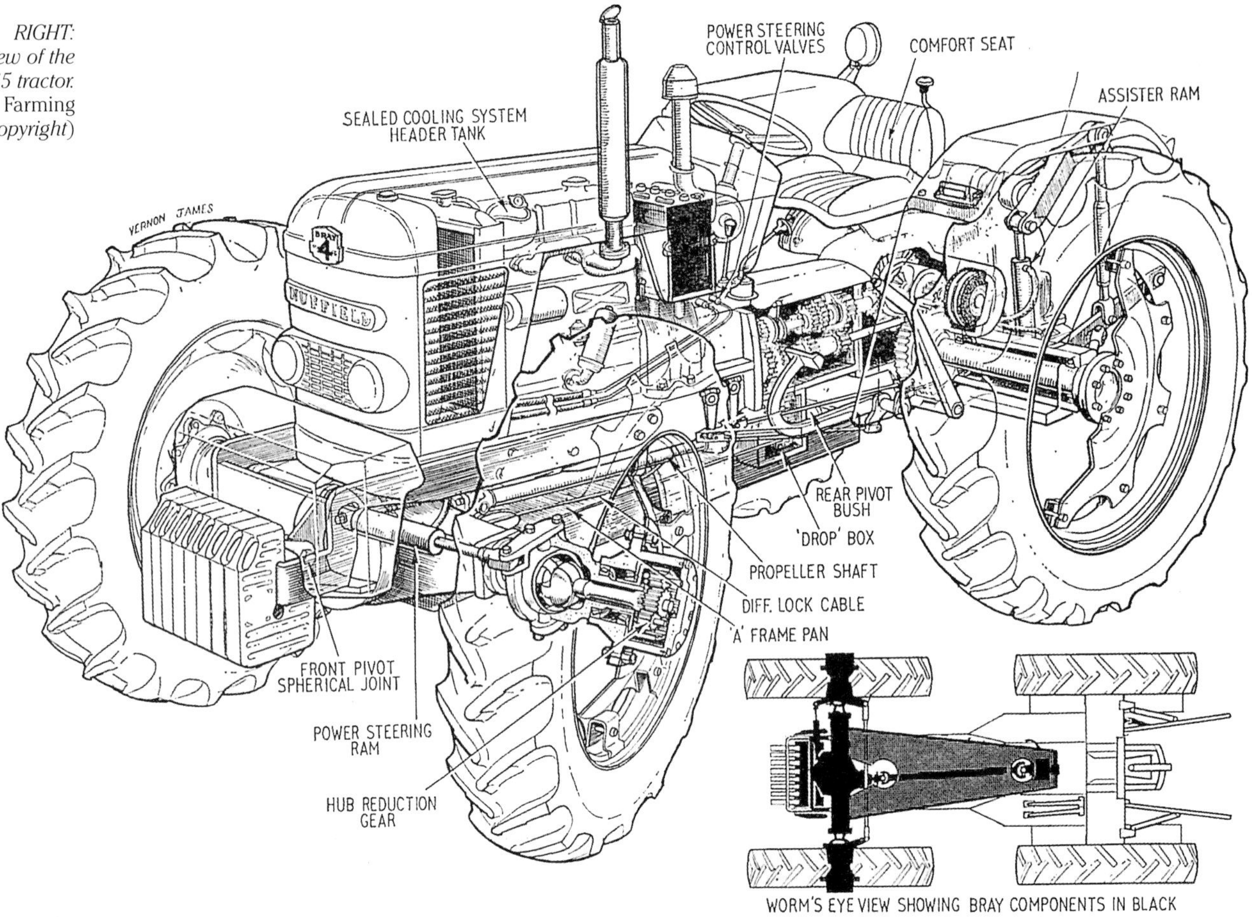

illogical not to mention the rest of the company's tractor production and complete the story.

In October 1965, layouts were drawn up for an agricultural and industrial Centaur II based on the Nuffield 10/60 tractor. The finished machine was released in March 1966 and called the Bray Four 10/60. It consisted of a basic Nuffield tractor fitted with a Bray front axle with planetary hub reduction. The front axle was powered by a centrally mounted propeller-shaft driven by a train of gears taken off the gearbox output shaft.

In May 1966, Bray built machine number AAK1254, a six-cylinder version, powered by a 5.1 litre BMC diesel. It was to have been known as the Bray 10/90, and aimed to compete with County's Super Six tractor, but it never went beyond the prototype stage. An industrial version of the four-cylinder model was also built.

With the introduction of the new model Nuffield 4/65, Bray showed its four-wheel drive Four 4/65 version of the 65 bhp machine at the Royal Show in July 1967, priced at £2,125. This model evolved eventually into the Bray Four 384, introduced in January

BELOW: The first experimental Matbro forklift, built in 1949, based on a petrol Fordson E27N tractor fitted with a second-hand mast.

LEFT:
Leonard Mathew at the wheel of the 1949 experimental forklift.

BELOW:
Another view of the 1949 experimental Matbro forklift. The counterbalance weights can be clearly seen under the tractor's front axle.

1970, based on the subsequent Leyland tractor. Bray also built a four-wheel drive version of the David Brown 1200 tractor in 1967, but this did not go in to production. Another venture by the company was the manufacture of a number of narrow conversion kits for the small BMC Nuffield 4/25 tractor, built until Leyland brought out its own version.

With the increasing property prices of the 1970s, Bray's premises at Feltham, close to Heathrow Airport, had become too valuable an asset. In March 1971, Bray were re-located to Tetbury in Gloucestershire, allowing the owners, Sheepbridge Engineering, to sell the Middlesex site. A new purpose-built factory of over 100,000 sq ft was built on a 10 acre site at Tetbury. Two years later, on 14 November 1973, Bray was sold to the Mathew family, owners of Matbro Limited.

The origins of Matbro date back to just after the end of the Second World War, when two brothers, Horace and Leonard Mathew, set up in business, at Wallington in Surrey, dealing in ex-War Department. equipment bought from the many war surplus auctions held in the late 1940s and early 1950s. H. & L. Mathew Limited was formed in 1946 as agricultural dealers and machinery suppliers, selling heavy electrical equipment, batteries, generators and compressors.

Most of the equipment bought and sold by the company had been disposed of by the US

RIGHT: The experimental Matbro forklift after being fitted with enclosed bodywork in September 1949.

Forces, and was of American manufacture, as was the brothers' second-hand Clark Stacatruc forklift truck. The problem with this machine was that it was designed for use on hard surfaces, and its small wheels were continually getting stuck in the brothers' unmetalled yard. This prompted Leonard, who was interested in engineering, to set about building his own machine with bigger wheels to cope with uneven surfaces.

His first experimental machine was built in 1949, and was based on a petrol Fordson E27N industrial tractor fitted with a second-hand Coventry Climax mast. The tractor's drive was reversed, and the steering wheel and driving position were turned to face the mast, and offset to aid visibility. Counterbalance weights were fitted under the fabricated steering axle. After successful trials, the machine was completed in September 1949 by the addition of enclosed bodywork.

Pleased with the final result, Leonard was keen to put the machine into full scale production. Horace reluctantly agreed, and forklift manufacture was established as an

ABOVE:
The forklift truck assembly line at the Matbro Works, Wallington, in 1952. The Fordson Major skid unit, used as the basis of the forklifts, can clearly be seen.

LEFT:
The Matbro Super Loadstar loading shovel introduced in 1953.

FACING PAGE - BOTTOM:
The first production model Matbro forklift, the Pneustac, at its press demonstration in May 1950.

nearly twenty years.

In 1951, Matbro produced the first British soft-ground forklift for military use, and a number were taken up by the army. The company also built a prototype small three-wheel forklift, fitted with the Ford 10 hp side-valve car engine, drive train and transmission, but as it was found to be unstable it was shelved. Following the introduction of the E1A Fordson Major in 1952, Matbro designed and exhibited the Series II forklift truck, based on this new skid unit.

The company's forklift trucks were built at the Matbro Works, situated in Sandy Lane North, Wallington, owned by Leonard and Horace Mathew's father, who leased it to his sons for a peppercorn rent. He had been an engineer in his own right, and in the 1930s had acquired the site, occupied by a wooden structure known as the Victory Hall which he had demolished and replaced with a new

ABOVE: The Matbro Super Loadstar photographed near the Eiffel Tower in Paris. The machines dominated the French market, and a number were built in a small factory just outside Paris.

off-shoot, using the Matbro name to differentiate the new product from the trading company.

The first production model forklift, known as the Matbro Pneustac, was demonstrated to the press by Leonard Mathew in May 1950. The machine was shown successfully negotiating soft grassland, pot-holes and a one-in-six gradient with a 30 cwt load without difficulty. Built around the Fordson E27N skid unit, it was available with a choice of petrol or Perkins P6 diesel engine, and single or twin drive wheels. The machine's lift capacity was 2 tons to a height of 9 ft, and electric starting was fitted as standard. The price was set at £950 for the petrol model, and £1,300 for the diesel version. This machine was to set the basic design for Matbro forklifts using the Ford skid unit for

RIGHT: Leonard Mathew tries out an experimental backhoe attachment fitted to a forklift skid unit in 1953. This was the machine that inspired the development of the Mantis.

10,000 sq ft factory, built with his own hands, brick by brick.

By 1952, the company was employing four assemblers who were turning out as many as nine forklifts per week off the production line. This was an incredible number of machines for such a small labour force, and admirably demonstrates how the use of a skid unit could simplify manufacture. In comparison, Matbro's production at Tetbury in 1991 required some sixty people to assemble twenty machines a week (admittedly these later machines were far more complex).

ABOVE:
The production model Matbro Mantis introduced in 1956.

The forklift trucks were further improved in 1958 with the introduction of the Series III machines, now incorporating the Power Major skid, torque converter drive and power steering. However, since 1952 the company had developed some other interesting products, in addition to the forklifts.

The first Matbro loading shovel had been introduced to the press on 18 May 1953, again based on the Fordson Major skid unit. Known as the Super Loadstar, the company claimed it was the first front-wheel drive loading shovel, although Bray had developed its HL27 machine at about the same time. Fitted with a 1 cu yd bucket, the Super Loadstar was capable of lifting 2,500 lb to a height of 10 ft,

LEFT:
The first Matbro Mastiff pivot-steer loading shovel was brought out in 1958.

RIGHT:
The Matbro 200 was the largest of the range of forklifts produced in the 1960s. Powered by the Ford 590E engine and fitted with a Brockhouse torque converter transmission, it could lift over nine tons.

BELOW:
The prototype Matbro Mastiff 6/100 MT agricultural tractor with the prototype fitter, George Marshall, at the wheel.

LEFT:
The prototype Matbro Mastiff agricultural tractor undergoing trials.

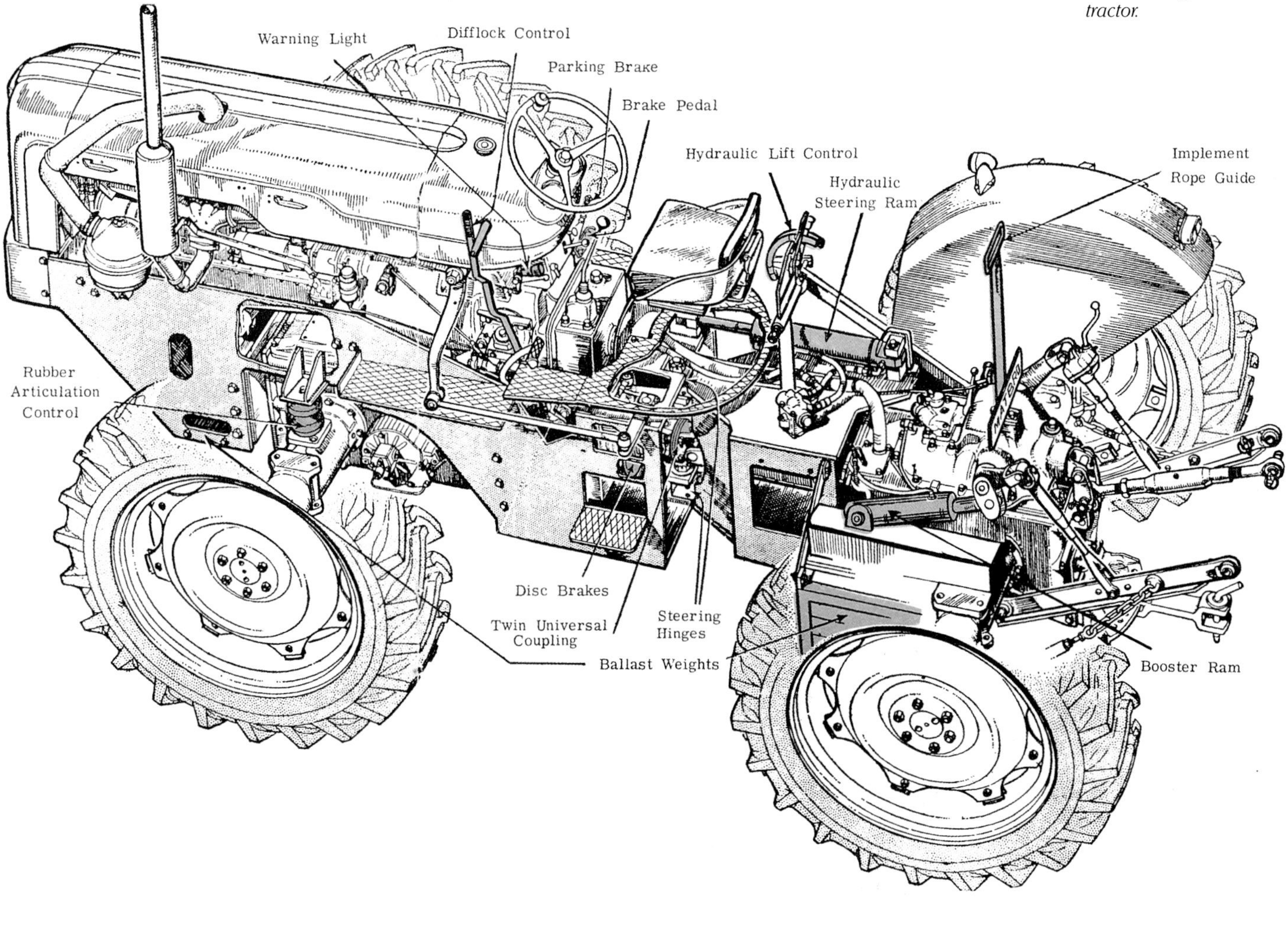

LEFT:
The Matbro Mastiff 6/100 MT agricultural tractor.

RIGHT: The production model Matbro agricultural tractor in 1963.

and proved to be a highly versatile and manoeuvrable machine which could load up to 5 tons of material in only four minutes. It was also available with forklift, dozer and crane attachments.

Unlike Bray's front-wheel drive loader, which was not so popular, the Matbro Super Loadstar became a very successful venture for the company, which experienced excellent export sales with it across most of Europe. In fact, the company totally dominated the French market with the machine, in sales to both France and her North African dependencies. Matbro's French distributor even set up a small factory outside Paris for limited production of the Super Loadstar. A four-wheel drive version was introduced in 1956, with moderate success, and production of the machine continued until 1964.

A further interesting diversification for the company came in the form of the Matbro Mantis, announced in October 1956. Derived from experiments carried out in 1953, when Leonard Mathew fitted one of the forklift trucks with a hydraulic backhoe in place of the mast, the Mantis was described as 'the first universal self-propelled hydraulic excavator.'

The production model Mantis was based on a similar front-wheel drive, rear-wheel steer unit to the Super Loadstar. It was fitted with a backhoe as well as a hydraulically operated bulldozer blade for back-filling trenches. Various other attachments were available, including a forklift, crane jib, winch, a

RIGHT: The Matbro Mastiff 6/100 MT tractor on demonstration. This machine is fitted with a Fritzmeyer cab.

high-lift operator's platform, and an air-compressor and receiver to drive further tools such as a concrete breaker. It was a remarkable machine, but was probably too ambitious for the time, and only about forty were made. One of the original Mantis machines is still owned by Leonard Mathew's son John.

Probably the most important aspect of Matbro's history was the development of the pivot-steer system. Leonard Mathew investigated the use of pivot-steer as a method of abolishing the 'wind-up' that occurred between the front and rear axles of normal four-wheel drive machines when turning on hard surfaces, causing transmission strain and power loss. If the machine was articulated in the middle, as in the pivot-steer system, both the front and back wheels would follow the same path when turning, and there would be no strain or 'wind-up' between the axles. Consequently, the machine could

BELOW: The Matbro Demonstration Unit, consisting of the tractor and a four-furrow plough.

LEFT: The Matbro Mastiff ploughing with a Ransomes Jumbotrac plough.

RIGHT: The Matbro Mastiff agricultural tractor on demonstration in 1963.

always operate under full power on any surface, with four-wheel drive permanently engaged.

The system had many other advantages - expensive drive-steer axles were not required, and no power was lost through complex steering joints. The simpler drive axles were also capable of carrying increased weight loads, and the turning circle was greatly reduced. Leonard Mathew started work on his 'centre-pivot steer' principle in 1955, and on 26 April 1957, together with his chief engineer, George Chapman, he applied for worldwide patents for the system, which were finally granted on 22 June 1960.

The first machine to incorporate the pivot-steer system was the Matbro Mastiff loading shovel, first shown at the Public Works Exhibition in 1958. The machine relied heavily on Ford tractor components. Two Fordson Power Major rear axle and transmission units were used to form the front and rear drive assemblies. The front unit was reversed, and the two input shafts were joined by a propeller shaft, fitted with a single universal joint at the centre pivot point. Drive was transmitted from a Power Major engine, mounted above the rear transmission unit, through a Brockhouse torque converter transmission and down to the propeller shaft by a transfer box.

Because of the need for increased production capacity, in 1959 Matbro moved to new premises in Horley in Surrey. The old works were eventually sold in 1961, and continued to be used as a factory until it was

BELOW: Matbro Mastiff serial number 1816, delivered June 1963, to dealers, Cleales of Saffron Walden

demolished at the end of the 1970s to make way for housing.

Also in 1959, Matbro, which until now had still operated as a subsidiary of H. & L. Mathew Limited, was incorporated as a separate limited company, with Leonard Mathew as chairman. Horace and Leonard each owned 50 per cent of the shares of both companies until the 1970s, when Horace sold his shares in Matbro in exchange for Leonard's shares in a third business, Power Plant Hire Limited, a plant and equipment hire

ABOVE:
Matbro Mastiff number 1816 after restoration.

LEFT:
The Matbro Mastiff Mk II 128 bhp agricultural tractor in 1967.

RIGHT: One of the Matbro Mastiff tractors with a trailing plough in 1963.

firm, which was managed by Horace until it was sold in the mid-1970s. The trading name of Power Plant Hire was retained by Horace, and is used today by his nephew, John Mathew, for his own hire company.

In the years that followed the opening of the new factory, a new line of forklifts was introduced, with lift capacities eventually ranging from 4,000 lb to 20,000 lb. The Mastiff Series II loading shovel was introduced in 1961, fitted with the more powerful Ford 590E six-cylinder engine and using Super Major transmission units.

Further advantages of the centre-pivot steer system had been discovered once the Mastiff machines had been proved in work. The capacity of the power steering to move the two halves of the chassis relative to each other gave the driver the ability to 'walk' the machine out of sticky patches when working in muddy conditions, and with the wheels following the same track, fewer ruts were formed in the ground. The driver could also use the power steering to slew the vehicle when stationary, useful when loading trucks, and it was not long before other manufacturers were showing interest in the system.

In 1962, to help recoup some of the development costs of the pivot steer system and generate its use on an international scale, Matbro concluded a $1 million licensing agreement with the American Caterpillar, International Harvester and Allis Chalmers companies, allowing them to incorporate the patented pivot steer system in their range of products.

1962 was also the year that Matbro built its first pivot-steer agricultural tractor. The Matbro Mastiff 6/100 MT was based on the same principle as the Mastiff Series II loading shovel, using two Super Major transmission units, but built in reverse, along the lines of a

BELOW: The 128 bhp Matbro Mastiff Mark II agricultural tractor outside the experimental workshop in 1967.

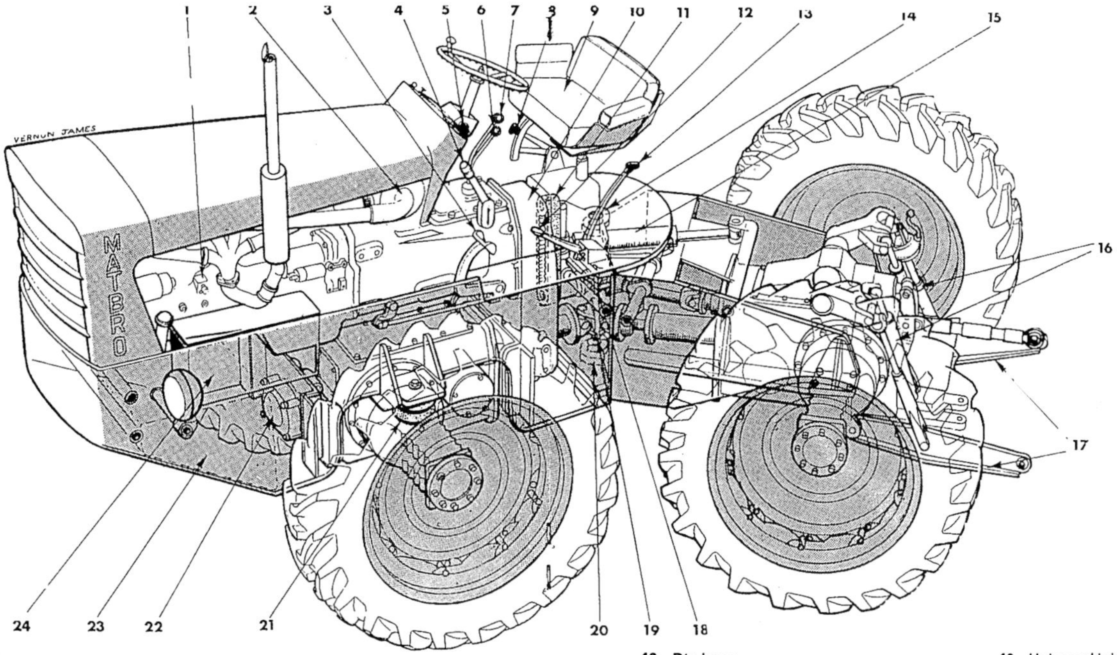

1 128hp diesel engine
2 Air cleaner
3 Clutch pedal
4 Hand brake
5 Steering wheel rake adjustment
6 High/low range gear lever
7 Main gear lever
8 Hydraulic control
9 Suspension seat
10 Transfer drive box
11 Toothed belt drive to pto
12 Diff. lock
13 Pto lever
14 Steering pivot point (upper)
15 Steering ram
16 Auxiliary lift rams
17 Reinforced lower links
18 Shaft to pto
19 Universal joint to main drive
20 Steering pivot point (lower)
21 Aeon cushion suspension
22 Front axle pivot
23 45-gallon fuel tank
24 12v battery

ABOVE: An exploded view of the Matbro Mastiff Mk II agricultural tractor. (Farmer and Stockbreeder *copyright*)

conventional tractor, with the driver placed in the usual position and the engine mounted over the front drive unit. Power from the six-cylinder 590E engine was transmitted through a standard Fordson six-speed gearbox to the transfer case, which housed a chain-drive to the propeller shaft. The machine also had the benefit of Fordson three-point linkage, powered by a high capacity front-mounted pump, and differential locks on all four wheels. It was finished in normal Ford tractor styling, with Super Major tin-work, cowl, bonnet and wings. The tractor was made at a new factory in Frome, Somerset, opened in 1963 for loading shovel assembly.

After development trials the tractor was demonstrated to the public at the end of 1962, and went on sale priced at £2,500. It was advertised as a 100 hp machine (which was rather optimistic for the 590E engine), and the company claimed it gave 'a crawler performance on wheels.'

The first tractor sold was delivered to Ford dealers Mann Eggerton of Ipswich, in March 1963. Eleven more were sold that year - eight on the home market, two to Sweden and one to Israel. Sales never materialised to the extent that Matbro would have liked, and the following year saw only eight more tractors delivered, three of which went to Denmark. At the end of the production run, only twenty Matbro Mastiff 6/100 MT tractors had been built.

The Mastiff performed well, and had excellent pulling power in the furrow. But it was expensive to buy, and the drive system did not allow for the fitment of a power take-off shaft, so the machine's use was limited to ploughing and heavy cultivation work.

The tractor also suffered some mechanical problems, which probably hampered further sales. The

BELOW: The Matbro Mini-Mastiff pivot-steer garden tractor.

replacement of the chain was a time-consuming task, normally requiring removal of the engine and gearbox. The second machine sold was modified by the owner, Mr. Watson of Corby Glen, Lincolnshire, to take a stronger Morse chain, and continued to work into the 1980s without further problems. This tractor, number 1755, is now in preservation, as is at least one other, number 1816.

Matbro made one further brief attempt to capture some farm tractor sales when it introduced the Mastiff Mark II agricultural tractor in 1967. This machine, powered by a 128 bhp version of the Ford 2704E engine, was fitted with a Ford 5000 gearbox which transmitted the drive though a transfer box to two 5000 rear axle and

ABOVE: Leonard Mathew pictured with the Matbro Mastiff Major loading shovel in about 1968. This machine used the Ford 2704E engine.

drive-chain in the transfer case proved to be a weak link in the transmission, and would sometimes break when the tractor was working under heavy load. Breakages normally occurred in the middle of harvest or during a busy ploughing season, and

transmission units, using a similar principle to the earlier tractor. The bodywork was built to Matbro's own styling, and housed a 45 gallon fuel tank mounted forward of the front axle, which when full, abolished the need for further ballast. The tractor weighed just over

RIGHT: A Goose elevating scraper hauled by a Muir-Hill 101 industrial tractor. Goose products were sold by Matbro after they acquired R. Smith & Son Ltd. in 1969.

LEFT:
The Matbro SLOD - swinglift load over driver - built from about 1973, and fitted with a Ford 2704E engine.

five tons, and was priced at £3,495.

Matbro had problems trying to source a suitable transfer box to provide the machine with a power take-off. With this in mind, several meetings were held with Roadless, but no agreement was reached. In the end, only one tractor was built, and although it was demonstrated all over the country, no orders were received.

The demonstration machine was eventually sold to Verolme, a large ship builder in Rotterdam, Holland. The tractor was heavily ballasted on the rear, and used to pull very large multi-wheeled trailers around the shipyard, carrying massive prefabricated ship sections. The gross weight of these trailers was reported to be sometimes as high as 400 tons.

The Matbro Mastiff agricultural tractor was probably too far ahead of its time. With a little more development, it could have been a very successful machine, but the company preferred to concentrate on industrial machinery sales. The market for pivot-steer agricultural tractors was effectively left open for Massey Ferguson to dominate during the 1970s, following the introduction of its 1200 articulated machine.

The company also made a small articulated garden tractor, known as the Matbro Mini-Mastiff, and billed as the 'World's Smallest Pivot Steer Tractor', with a choice of 12 hp petrol or 7 hp diesel engines. Unfortunately, it was no more successful than its larger counterpart, and only one was built, which gave stalwart service at Leonard

LEFT:
The largest of the Mastiff loading shovels, the 1000 model, was usually sold under the Bray name after 1973, but occasionally they carried Matbro transfers for certain export markets. This machine was destined for use in the pulp industry of southern Turkey in 1974.

TOP: The Bray AG 4400 agricultural pivot steer loader, produced from 1979, and powered by a four-cylinder Ford 2712E engine developing 80 bhp.

ABOVE: The Matbro RAM 40 radial arm farm handler.

Mathew's nursery! One Matbro industrial tractor was also produced, and was basically the Mastiff loading shovel without the loader arms fitted. It was sold to the British Army for towing heavy-duty trailers on a camp near Aldershot.

During the 1960s, the Matbro product range was consolidated, and new machines were developed. The Matbro Swinglift was introduced in 1964, designed as a dual-purpose handler. Unlike a conventional forklift, this patented machine was capable of slewing its load through 90 degrees and carrying it parallel to the direction of travel.

Eventually new models of the loading shovels were introduced, including a smaller version, the Mastiff Minor, fitted with the four-cylinder Ford 5000 engine. The Series III machines, including the Mastiff Major 125, powered by a Ford 2704E engine, used 5000 rear axle and transmission units, and were the last Matbro shovels to source their drive-axles from Ford skid units. All subsequent machines used unitary components for the axles, but the company continued to use Ford engines.

Another new factory was opened in 1965 at Dunmurry, Northern Ireland, to manufacture electric forklift trucks. Further expansion was implemented in 1969, when Matbro acquired the firm of R. Smith & Son (Horley) Limited, a structural steel manufacturing company located directly opposite the Matbro works in Horley. R. Smith & Son had diversified into manufacturing rear-dump semi-trailers and elevating scrapers, made under licence from the Rhodesian Morewear company, and sold under the 'Goose' name. The name for the dumpers was derived from the 'goose-neck' coupling designed to transfer the trailer payload on to the tractor's driven wheels. Matbro continued with production of the dump trailers and elevating scrapers for several years.

By 1973, Matbro had a product range that included over twenty forklift trucks, eight models of swinglift, five rough terrain forklift trucks, four Mastiff loading shovels, two dumpers and a scraper. Following the acquisition of Bray in that year, the production of Matbro wheel loaders was transferred from Frome to Tetbury, along with the rough terrain forklifts, while Horley continued to make the industrial forklifts and the Swinglift.

The Bray and Matbro loading shovel ranges were integrated, and new designs were intro-

LEFT:
The Bray T 100 experimental pivot-steer for industrial and forestry use, built in 1980.

BELOW:
The experimental Bray D120 articulated 12 ton dump truck.

duced under chief engineer, Eddie Stevens. The Matbro pivot-steer concept was adopted for all the four-wheel drive machines; a number of two-wheel drive, rear-wheel steer loaders were also produced.

For uniformity of marketing, the Bray name was retained for all loader production, as it was felt to be stronger and better known in the market than Matbro's, although the latter's products were more technologically advanced. For several years, Bray Construction Machinery Limited was to continue as a separate concern, run by Leonard Mathew's son John, and during the late 1970s, following large orders from Algeria and Columbia, production of Bray loaders peaked at 400 units per year.

Sadly, Matbro's founder, Leonard Mathew, was unable to involve himself with the newly expanded company for more than a few years, and died in August 1981, aged only sixty-nine. Ironically, his brother, Horace, passed away the same year. Following his father's death, John Mathew merged Bray into Matbro Limited, and took over as chairman of the company. The trading name of Matbro-Bray was used until the end of the 1980s.

In the late 1970s, Matbro moved into the field of handling specialist agricultural materials with the introduction of the Ram 40 (Ram standing for radial arm machine). The Ram 40 was a multi-purpose, four-wheel drive mechanical handler, with a single lifting arm that could be fitted with a number of attachments, mounted on a pivot-steer chassis. The concept was taken one step further when Matbro brought out the Teleram, a pivot-steer handler with a telescopic boom, at the 1981 Smithfield Show. A rigid chassis version of the telescopic loader, the Telestar, was intro-

RIGHT: The front yard of Matbro's Horley works in the early 1970s. Leonard Mathew is standing on the Mastiff loading shovel at the rear.

duced at the 1983 Royal Show, and the four-wheel drive model of the same machine made its debut at the following years event. A non-telescopic rigid chassis loader, called the Loadstar, was also brought out. All these machines used either Ford or Perkins engines.

The Ram 40 was based on the same pivot-steer chassis as the rough terrain forklift, and Matbro felt that the same basic space-frame chassis could be used for a number of other machines. By reversing the driving position, the company was able to build a prototype tractor for industrial and forestry work. Brought out in 1980, under the Bray name, the machine was designated the T100, and fitted with a 114 bhp Ford 2714E six-cylinder engine. Matbro tried to persuade the Forestry Commission to buy the tractor, but lost out to more fully developed, but dearer, Swedish machines.

The company then experimented with using the T100 with a Goose dump trailer, before taking the concept one step further and fitting the same front half of the machine to an extended rear unit, fitted with a dump body, to make an articulated four-wheel drive dump truck. It was called the D120, standing for dumper with a twelve ton payload, but never went beyond the experimental stage. In some ways, this machine was the forerunner of today's Caterpillar, Moxy and JCB articulated dump trucks.

Through the 1980s, Matbro rationalised its product range, concentrating on the telescopic handlers and a limited range of loading shovels. On 31 March 1989, forklift manufacture ceased, and all production was transferred to Tetbury. The Horley premises still

BELOW: The Bray experimental T100 tractor fitted with a Goose dumper.

belong to John Mathew, and are leased out to unrelated companies.

The Teleshift, a telescopic twin-boom materials handler, was added to the Matbro range in 1990. Originally designed by Ron and Tony Collins of RWC (Developments) Limited, in the late 1970s, it was manufactured and sold under the Roadless name for a time. RWC were acquired by the agricultural and industrial division of the Benson Group in 1985, and the Teleshift was sold under the Benson name before eventually being bought by Matbro.

In July 1991, Matbro was bought by Powerscreen International, a British company based in Northern Ireland. Since the acquisition, Matbro's output has increased nearly ten-fold, and with the production line at Tetbury running at full capacity, the company have opened up a new factory in Northern Ireland. Today's range consists of three loading shovels, five telescopic handlers and limited production of the Teleshift.

ABOVE:
One of the current Matbro range, the TR250 pivot-steer telescopic handler has a lift capacity of 2.5 tons and is powered by a Perkins four-cylinder turbocharged engine developing 106 hp.

LEFT:
165 hp John Deere engine powers the current Matbro LS300 agricultural loading shovel seen here on maize silage work.

CHAPTER 7

Other Conversions

A new tractor is delivered, and the old machine is loaded on the lorry ready to go away. Both Fordson E27N Majors are fitted with Rotaped tracks and Muledozers.

ABOVE: A narrow version of the Fordson Roadless crawler, pictured working in a Kent orchard with a nine-tine cultivator in 1931.

It would be impossible to write a book about Ford tractor conversions without giving the Roadless company a brief mention.

From 1923 under the motivation and guidance of the founder, Lt.-Colonel Philip Johnson, Roadless Traction Limited was a magnificent forward-thinking company pioneering the development of off-road transport in all forms, from wheelbarrows and bullock carts to modern four-wheel drive tractors of over 100 hp. In the early days the Fordson was only one machine among the many various makes and types of tractors and vehicles that Roadless converted on to tracks, but as the company progressed, it began to rely more and more on the Ford as a basis for its products. In the later years of four-wheel drive production, the use of Ford tractors by the company became predominant, but even then a few four-wheel drive conversions on other makes crept in.

The first Roadless conversion of a Fordson tractor dated back to 1930. Ford dealers Invicta Motor Engineering Works of Canterbury were invited by Margate Corporation to tender for the supply of a tractor to haul seaweed off the beach, a task previously undertaken by horses. As a normal wheeled tractor was found to be unsuitable for pulling the wagons heavily laden with wet seaweed on loose sand, Invicta Motors approached Roadless to build them a Fordson with the rubber-jointed tracks that the Hounslow-based company had previously successfully fitted to Rushton and other makes of tractor.

Plans were drawn up in late 1929, and the first Fordson-Roadless tractor was born. The machine had its initial demonstration in March 1930, with favourable results, easily hauling a ten ton load of gravel on wet sand, even when the trailer wheels sank to a depth of 9 in.

The Fordson was such a well-known machine that Roadless was soon receiving orders and enquiries for the tracked version

ABOVE:
An early Roadless four-wheel drive conversion of a 1956 Fordson Diesel Major.

LEFT:
Roadless built many four-wheel drive conversions on Ford tractors. This one-off conversion of the American 8000 was carried out for Ford France SA.

ABOVE:
A Ford 7600 fitted with a Schindler four-wheel drive conversion supplied by Farm Tractor Drives of Derby.

RIGHT:
A non-Ford use of the Schindler axle is illustrated by this conversion carried out on a Massey Ferguson 290.

from all over the world. Within a few months, Fordson-Roadless tractors were operating in Spain and Turkey. In January 1931, the Ford Motor Company sanctioned approval of the Roadless conversion, and agreed to market the tractor through its dealer network. Prices were set at £350 for the standard machine, and £385 for an extended version with an extra bottom roller.

This was the start of Roadless's long association with Ford tractors. Full-track and half-track versions of Fordson tractors were produced through the 1930s and 1940s, many of which were used extensively on airfields during the Second World War. Full-track and DG half-track versions of the E27N and the newer model Fordson Major were also built.

Roadless was introduced to the four-wheel drive market by chance. Following the introduction of the new Major, second-hand tractor importers Selene of Turin were faced with stock of old-type Fordson E27Ns on their hands. The Italian company decided to encourage sales for these obsolete models by converting them to four-wheel drive. This was achieved by fitting war-surplus American military vehicle front

ABOVE:
The prototype FTD Chieftain tractor, fitted with the Schindler axle and powered by the Ford 2715E engine.

axles, stocks for which had been left all over Europe after the Second World War, and taking the drive off a transfer box mounted between the tractor gearbox and the differential housings.

The tractors sold well, and in December later that year, Signor Segre, head of the Selene firm, met with Sir William Batty of the Ford Motor Company at the Smithfield Show to discuss future possibilities. Ford did not want to get involved with the conversion, but agreed to continue to guarantee any tractors fitted with Selene's four-wheel drive system. One man who was very interested in the conversion was Lt.-Colonel Johnson of Roadless.

Johnson encouraged Selene to develop the idea, and in 1953, arranged for one of the four-wheel drive Majors to be tested by the NIAE at Silsoe, where it received a very favourable report. It was natural, therefore, when Selene wanted to appoint someone to manufacture and market the conversion in the UK, that they should approach Roadless Traction. In 1955, Roadless were granted the patents licence to manufacture and market the Selene conversion for Ford tractors throughout the British Commonwealth. Manufacture started at Hounslow the following year, heralding the beginning of Roadless's production of four-wheel drive Ford tractors, which was to continue into the mid-1980s. Roadless eventually developed its own four-wheel drive axle, and also built equal-size wheel tractors in the form of the 115 and 120 models.

Selene continued to develop further 4 x 4 conversions for the Ford, which were mainly supplied as kits, as well as manufacturing four-wheel drive systems for other makes of tractor. By the end of the 1960s, the company was marketing the 4RU, an equal-size four-wheel drive conversion for the Ford 5000.

Many European-built Ford tractors were fitted with Selene axles. In Holland, they were marketed by Ford of the Netherlands and

RIGHT:
A Dutch Ford 7000 fitted with a Selene four-wheel drive front axle.

fitted to most of the four-wheel drive Ford tractors sold in that country during the 1970s. Selene axles were also used in Belgium, and Ford France fitted them to their six-cylinder 5095 tractor. The 5095 tractor was also a conversion in itself, built up for Ford France by EVA, Belgian importers of Ford industrial engines, using the 95 hp 2704E engine.

Selene was bought out at the end of the 1970s by the Swiss company, Schindler. Established in 1874 as a manufacturer of rolling stock for the railway, Schindler diver-

RIGHT:
The Selene axle fitted to a Belgian-built Ford 7600.

LEFT:
Ford 6710 Force II tractor, fitted with a Schindler four-wheel drive front axle, photographed in Holland.

LEFT:
The third FTD Chieftain tractor to be built. This machine spent its working life on the Isle os Sheppey.

sified into the agricultural market in 1965, producing four-wheel drive conversions for tractors.

Farm Tractor Drives, run by Arthur Battelle in Derbyshire, were the UK agents for Schindler four-wheel drive conversion kits from 1977 to the mid-1980s, when the agency went to Harris Agricultural Engineers for a short time. Schindler built kits to suit several different makes of tractor. Farm Tractor Drives offered conversions for both Ford and Massey Ferguson tractors, but over 90 per cent of the kits sold went on Ford tractors. Two types of axles were available - the EL65 for tractors up to 70 hp, and the heavier PLK for larger machines. Drive for the axle was taken off the gearbox by a side-mounted transfer box.

At the 1982 Smithfield Show, Farm Tractor Drives released details of its Project 82, a move by the company into building a complete tractor based on an uprated Ford 5000 transmission fitted with a more powerful engine and the Schindler axle.

Designated the FTD Chieftain, it was designed as a low-cost, rugged and

BELOW:
The first FTD Chieftain tractor on Farm Tractor Drives' stand at an agricultural show.

RIGHT: An American export model of the British-built Fordson Super Major, transformed into an industrial loader and fitted with Levy hydraulic Live-Power four-wheel drive.

RIGHT: The Carron Hytorque hydrostatic four-wheel drive conversion for the Ford 5000.

BELOW: HRH The Duke of Edinburgh tries out the NIAE experimental hydrostatic tractor at Silsoe.

economical tractor. It was fitted with a Duncan de-Luxe cab, and initially priced at £13,750. The first prototype was built in 1983, fitted with the Ford 2715E engine. Only two others were made, both with 2725 engines developing 105 hp, although two further machines were built up by customers using parts supplied by Farm Tractor Drives. These farm-built machines were known as the FTD Warriors.

Ford also used Schindler axles on a couple

of its tractors. Axles were supplied directly from Schindler to Ford at Basildon for the 8200 tractor, introduced in 1980. The previous model, 8100 tractor, was not fitted with four-wheel drive in Britain, whereas the French version, built by EVA, had the option of the Schindler axle.

In about 1985, Schindler moved out of the four-wheel drive conversion market to concentrate on its other businesses, which included escalators and lifts as well as railway rolling stock. Other companies, offering four-wheel drive conversions for Ford tractors included the Italian companies Palazanni and Carraro. Carraro axles were eventually factory-fitted as standard on Ford tractors, along with the German ZF four-wheel drive axle, although from 1 January 1995, New Holland Ford used only its own Fiat-built axles.

Apart from the various mechanical four-wheel drive conversions built

ABOVE: Ford 5000, fitted with Carron Hytorque four-wheel drive on demonstration.

LEFT: The Shire 400 industrial tractor, based on the Ford 4000, built by Perry's of Potters Bar.

RIGHT: The Shire 500 industrial version of the Ford 5000.

for Ford tractors, a couple of companies developed systems using hydraulic wheel motors to power the front wheels. The NIAE pioneered the use of such motors with an experimental hydrostatic machine, based on the Fordson Major, built in 1954 by the Mechanical Engineering Department. Designed by H. J. Hamblin, the tractor was driven by hydraulic motors in the rear wheels. Speed and direction of travel were determined by a lever which controlled the flow of oil from a Vickers variable-speed gear pump, chain driven off the Fordson engine crankshaft.

One of the first hydraulic four-wheel drive systems to be offered for use on Ford tractors

RIGHT: Perry's Shire 500 industrial tractor, based on the Ford 5000 and fitted with a Carron Hytorque hydrostatic four-wheel drive axle in 1972.

LEFT:
Fordson Super Dexta and Dexta tractors fitted with high clearance kits made by Gates of Baldock.

BELOW:
A narrow orchard conversion of an early Dagenham-built Fordson N made by Stanhay of Ashford, Kent.

was built and designed by Levy Industries Limited of Toronto, Canada. Levy's system was developed for the industrial Fordson Super Major fitted with the American 712 loader. A front-mounted pump, driven off the crankshaft, provided power to hydraulic motors positioned inside the front wheels. Levy also offered the hydraulic four-wheel drive conversion for other makes of tractor.

A British company, Carron Hydraulics Limited, also developed a hydrostatic system of four-wheel drive for Ford tractors in 1971. The Carron Hytorque four-wheel drive conversion for the Ford 5000 used two axial high torque, low speed hydraulic motors fitted directly into the front wheels, which were mounted on replacement stub axles. The motors were powered directly from an engine-driven gear pump, and an advantage of the system was no loss of under-axle clearance. In practice the motors ran hot and proved unreliable; consequently, Ford would not recommend the conversion or supply skid units to be fitted with it. Roadless also used Carron hydraulic motors for the drive and winch power on some of its early Logmaster pivot-steer forestry tractors.

The Carron Hytorque four-wheel drive conversion was marketed for a while by the Ford tractor dealers Perry's of Potters Bar, Hertfordshire, and was offered as an option on their Shire industrial tractors. Perry's Shire 400 and 500 tractors were highway and industrial versions of the Ford 4000 and 5000, altered to comply with road traffic regulations and fitted with hydraulic disc brakes and heavy-duty towing hitches. They were built in the mid-1970s to fill a gap in the market at a time when Ford were only marketing industrial and highway versions of the 2000 and 3000 tractors.

Many Ford tractor dealers built their own conversions to meet the special needs of their particular areas. Lincolnshire dealers Bettinsons of Holbeach built a single front-wheel conversion for the Fordson Major to suit the rowcrop work of the Fens, while Gates of Baldock marketed a high-clearance version of the Dexta for the vegetable and salad growers of Bedfordshire and Hertfordshire. In Kent, the orchard and hop farms were catered for by a narrow-gauge version of the Fordson N, built during the 1930s and 1940s by Stanhay of Ashford.

Stanhay also built a number of cranes fitted to the industrial Fordson tractor. The cranes were operated by a piston worked by compressed air provided by a belt-driven compressor. An air storage tank was mounted at the rear of the tractor.

Following the introduction of the new Fordson Major, production of narrow tractors in Kent was taken over by Stormont Engineering of Tunbridge Wells. Stormont's narrow gauge conversions of the Major were built at their Hillsborough workshops, and were sold under the KFD name, standing for

BELOW: Advertisement for the Bettinson three-wheel conversion for the Fordson Major.

LEFT:
The KFD 68, a low orchard conversion of the Fordson Major, is demonstrated outside the works of Stormont Engineering.

LEFT:
This photograph of the KFD 52 narrow tractor standing on Stormont's forecourt shows its height in comparison with that of the old Standard Fordson.

BELOW:
KFD 64 industrial tractors working in Ford's Dagenham plant.

Kent Ford Dealers. Prototypes were built up in June 1952, and the KFD was available from March 1953, initially in two versions, the 52 narrow gauge machine, and the 68 orchard tractor, with overall widths of 54 and 68 inches respectively. Stormont also made a hop plough, tiller and a seven-tine cultivator to match the tractors.

In 1955, Stormont introduced the KFD 64, a narrow industrial tractor. Several of these machines were used by the Ford Motor Company at its Dagenham factory to move materials around the 560 acre site. Many KFD narrow gauge tractors were exported for use by growers of sisal, sugar, coffee, olives and citrus fruits, and it continued in production through to the Super Major.

Apart from the County and Roadless

RIGHT:
The Stanhay portable crane was introduced in 1942, based on an industrial Fordson, and operated by compressed air.

RIGHT:
Mailam 5001 crawlers lined up outside the Venice works in 1966.

ABOVE:
Bulldozer version of the Mailam 5001 crawler, based on the Ford 5000 skid unit.

LEFT:
Ford 7610 fitted with experimental rubber half-tracks developed by Silsoe Research in 1991.

crawlers, few other tracked versions of Ford tractors were produced. A crawler based on the early Model N Fordson was built in Scotland for a short time in the 1930s, but only a very few were produced. When production of the Super Major came to an end at Dagenham, both County and Roadless dropped their crawlers, and it was left to Mailam of Italy to produce a tracked version of the new Ford 5000. County had approached Mailam to build them a crawler based on the 5000, but when agreement could not be reached, Mailam brought out its own version.

Mailam was founded near Venice in the late 1950s. The founder and managing director, Dr Segattini, had combined his experience in Caterpillar machines with that of his chief engineer, who had worked with International Harvester crawlers, to produce two industrial crawler tractors. The 55 bhp 655, and the 67 bhp 670 machines were both available as loaders or bulldozers. The 670 was particularly advanced for the time, having a twin-disc torque converter and power-shift transmission.

In 1966, the company brought out the 5001 agricultural crawler, based on the Ford 5000. Steering was by multiple disc clutches, and it was fitted with Berco track running gear. A six-cylinder version soon followed, powered by the Ford 2703E engine. In 1968, the four-cylinder 5001 was uprated to 75 bhp, following the change to Ford Force skid units, and the six-cylinder model was fitted with the 2703C engine giving 115 bhp. Industrial versions were available with a bulldozer blade.

Ford agents Gates of Baldock gained the British concession for the 5001 in 1970, and demonstrated it for the first time in October at a ploughing match in Barkway, Hertfordshire. The price was set at £3,995 for the four-cylinder tractor, and £5,250 for the six-cylinder machine. In the event, Gates only managed to sell two of the four-cylinder models, and only one of the more powerful six-cylinder tractors.

One further Ford-based crawler was built in

BELOW: Six-cylinder version of the Mailam 5001 crawler, powered by the 2703C engine.

the 1980s, by Intermec Farm Machines of Calne, Wiltshire. This small machine was based on the Ford 1210, and was designed for the horticultural user.

Several half-track conversions were available for Ford tractors. Apart from the Roadless DG girder tracks, flexible half-track kits were built by the Canadian Bombardier and the American Arps companies. Similar systems were marketed in Britain by JCB and County. Even in 1991, Silsoe Research built an experimental Ford 7610 fitted with rubber half-tracks which allowed ploughing to be carried out in the furrow with a 30 per cent increase in tractive performance.

One of the most interesting half-track conversions to be used on the Fordson tractor was the Rotaped. Unlike other half-tracks, Rotaped equipment consisted of six jointed pans forming a self-laying track, kept in formation by tensioned chains. The drive was provided by a large sprocket wheel bolted to the tractor hub in place of the conventional rear wheel. The advantage of the system was that conversion from wheels to Rotapeds only took about thirty minutes.

Invented in 1941 by O. Singer of London, the tracks were originally manufactured by George Monro Limited of Ingos Works, Waltham Cross, Hertfordshire, and sold as the Ingos Rotapeds. Singer went on, in 1946,

ABOVE: The Rotaped track unit.

BELOW: Fordson E27N Major fitted with Rotapeds, Muledozer and a Holman Tractair compressor.

RIGHT:
The Monro Muledozer fitted to a Fordson E27N Major tractor.

BELOW:
The Muledozer is raised and lowered by a screwjack operated by the wheel positioned next to the driver.

LEFT:
Leeford's improved Rotapeds fitted to a Fordson Diesel Major. The tracks were kept in formation by the pulleys running inside the steel rail.

LEFT:
Fordson Diesel Major fitted with Rotapeds and a hydraulically operated Muledozer.

ABOVE:
Fordson Dexta, fitted with Leeford Rotapeds, hauling a set of disc harrows.

RIGHT:
The Leeford hydraulic Muledozer fitted to a Fordson Diesel Major. The front-mounted oil pump is clearly visible.

ABOVE:
Leefords also equipped other makes of tractor - as illustrated by this International B250 fitted with Rotapeds and a hydraulic muledozer.

to invent a rubber track, similar in design to the system used on today's Caterpillar rubber tracked crawlers, and predating it by over forty years.

The first public demonstration of a Fordson fitted with Rotaped tracks was held near Harlow, Essex, in April 1942, with the tractor pulling a four-furrow Massey Harris No. 26 plough. Monro claimed that the track equipment increased drawbar pull by over 100 per cent, and reduced ground pressure by 85 per cent to 7 psi. By 1944, the tractor was also fitted with independent brakes to aid turning.

By the mid-1940s, Monro was building Rotapeds to fit the E27N Major, and in 1946, the company introduced the Muledozer. The Muledozer was a bulldozer attachment mounted on the front of the tractor, the blade raised and lowered by a screwjack operated by a wheel mounted close to the driver's seat.

Manufacture of the Rotaped track equipment and the Muledozer was taken over in the 1950s by Leeford (London) Limited. Leeford introduced cast alloy track plates, and for a time experimented with a system of pulleys running inside a rigid steel rail to keep the track in formation, before returning to the original chain method. The company also developed a hydraulically operated Muledozer fitted with twin rams powered by a front-mounted pump driven off the crankshaft. Leeford continued to offer Rotapeds, even for the Ford 5000, up to about 1967, but the hydraulic Muledozer became a more popular product, and several were fitted to Roadless and County crawlers.

Rotary Hoes Limited, of Horndon in Essex, which based many of its Howard products on Ford tractors, was also a regular user of Rotaped tracks. Better known for its range of Howard Rotavators, the company also produced drainage machines and several other interesting conversions on Ford tractors.

During the war, when paraffin fuel was in short supply, Rotary Hoes built a portable gas producer plant attachment for the Fordson. The producer plant worked by drawing steam through glowing coke or anthracite to form a gas which the engine would run on. The heavy plant, weighing some 5 cwt, was mounted on

ABOVE:
Fordson New Performance Super Major fitted with a Leeford Hydraulic Muledozer.

RIGHT:
1940 Fordson tractor fitted with a Howard Gas Producer.

LEFT:
The Howard Dungledozer manure loader based on the Fordson tractor.

BELOW:
Rear view of the Dungledozer showing the rotors for shredding the manure.

BELOW: The Howard gas producer plant was mounted on the nearside of the tractor to aid weight distribution when ploughing.

the left-hand side of the tractor, the idea being that it would be counter-balanced by the weight of the tractor when ploughing with the offside wheels in the furrow.

The producer gas plant was not a great success. Although only 10 lb of anthracite was needed to provide enough fuel to equal one gallon of paraffin, power output was down by 40 per cent. While the fire in the grate did not need as much attention as that of a steam engine, it was another job the tractor driver could do without, and heat from the plant could make his lot even more uncomfortable on a hot day.

The Howard Dungledozer was another short-lived machine. Designed, as the name suggests, to load dung or manure from a crew-yard, it was based on a Fordson tractor fitted with Rotaped tracks and a reduction gearbox. The machine was reversed slowly into the manure heap, allowing a pair of rear-mounted rotor blades to shred the

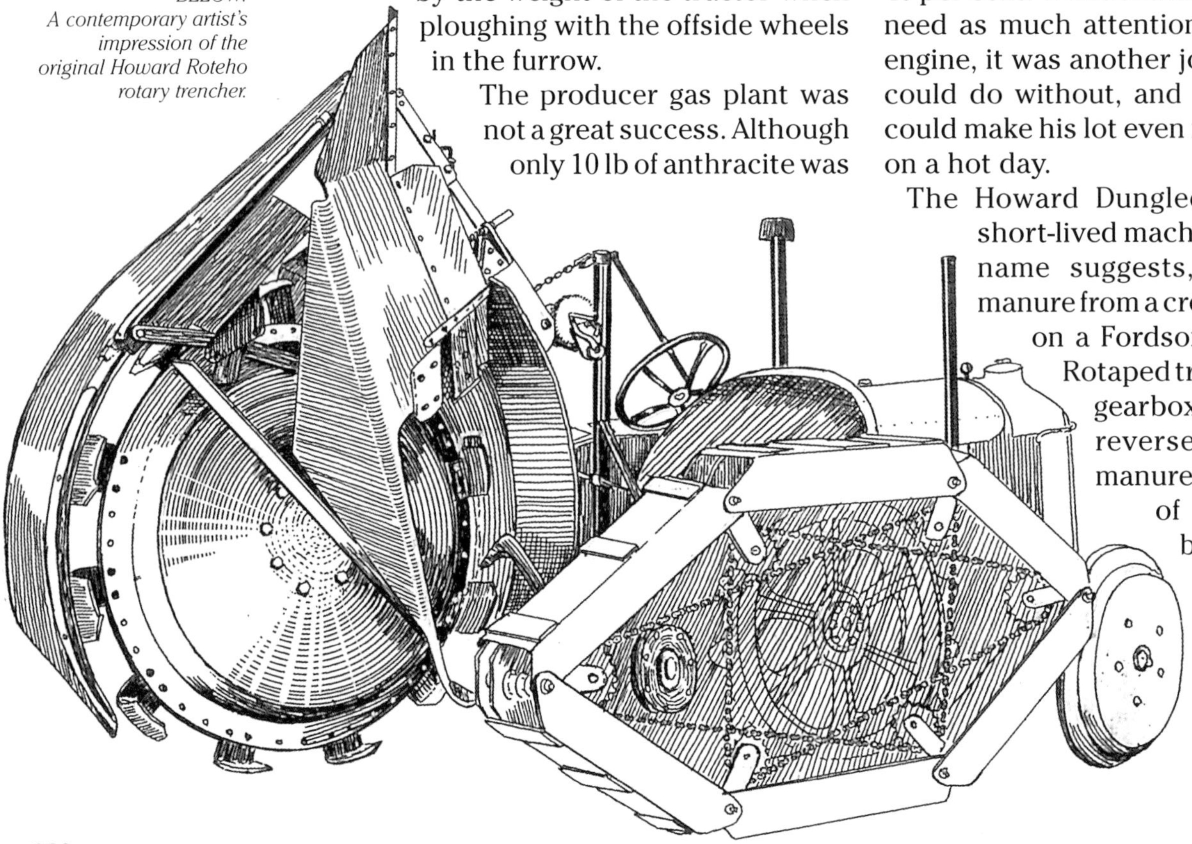

BELOW: A contemporary artist's impression of the original Howard Roteho rotary trencher.

ABOVE:
This Howard trenchdigger fitted to the new Fordson Major was introduced at the 1952 Smithfield Show.

LEFT:
Allen drainage machine built during the 1960s, fitted with conventional tracks and powered by the six-cylinder Ford 590E engine.

RIGHT: The Barth K140 Mk 4 trencher powered by the 75 hp Ford 5000 engine.

BELOW: The Whitlock Dozaloader fitted to a Roadless four-wheel drive Super Major. This tractor has the industrial axle with planetary hub reduction, introduced in 1963.

manure and feed it up an elevator into a trailer. The machine was efficient and would load up to 25 tons of manure an hour, but was cumbersome and was quickly outdated by the arrival of the front-end loader.

Rotary Hoes had more success with its drainage machines. The original Roteho trench digger and pipe layer was sold as an attachment for the Fordson during the 1940s. The tractor was fitted with Rotapeds and had a reduction drive taken off the worm shaft. A heavy roller chain, driven off a bevel gear on the pto shaft, transmitted power through a spur reduction gear to the trench digging rotor. The rotor was fitted with twelve blades, and was raised out of work by a hand winch. Drain pipes were laid by a tile-layer mounted on the rotor blade. With the arrival of the E27N Major, the company introduced the Howard reduction gearbox which was fitted between the gearbox and the rear axle transmission to give better creep speeds.

LEFT:
On this Allen machine, the digging rotor was raised and lowered hydraulically. Remote driving controls can be seen on the platform, left of the rotor.

BELOW:
The Howard Trench Digger Mk IV, based on the Ford 5000 fitted with Rotapeds and seen working in 1968. The driver is controlling the machine remotely from the rear.

BELOW:
The Allen Landrainer Mk 4. Powered by a Ford 5000 skid unit, it was made by John Allen and Sons (Oxford) Limited.

A new Howard Trench Digger was shown at the 1952 Smithfield Show. Mounted on a new Fordson Major fitted with Rotaped tracks, and supplied as a complete unit, the improved trencher was capable of trench digging and laying tiles at the rate of 175 yd an hour.

Further modifications were made over the years, including the provision of hydraulic rams to raise and lower the digging rotor and remote control steering and stopping. Many of the Fordson Major tractor units supplied

RIGHT:
The Barth K140 Mk 3 trencher was available with either Ford 5000 or 7000 power unit.

RIGHT:
The McConnel-Thornton-Garnett TG3 trenchless drainer fitted to a Roadless 115 tractor.

ABOVE:
Fordson Major on Rotapeds fitted with a SkyHi front-end loader.

LEFT:
An American Ford 6000 tractor fitted with a bulldozer attachment built by Waldon of Fairview, Oklahoma. Built in the early 1960s, the 6000 was fitted with a six-cylinder diesel engine.

RIGHT: Early Howard Roteho trench digger based on the Fordson N.

BELOW: A Allen Landrainer, built by John Allen of Oxford, using the Ford 5000 power unit.

ABOVE:
An American Waldon hydraulic angledozer attachment seen on a Ford FW20 in 1977. The FW20, the smallest model in the range of high horsepower articulated tractors built for Ford by Steiger, was not brought into the UK.

LEFT:
An Australian Malcolm Moore Roadmaster Major grader. Originally based on the E27N Major, this machine has been extensively modified and now has a Ford 2712E engine.

FACING PAGE: Cameron Gardner Rearloda fitted to a Fordson Major tractor.

with Howard trenchers were fitted with Perkins 4.270 engines by Ernest Doe & Sons to give the machines more power. In the 1960s, Howard produced some larger drainage machines fitted with conventional tracks and Ford 590E engines.

The last version of the Howard Trench Digger, the Mark IV, was introduced in about 1967, based on the Ford 5000 and fitted with Rotaped tracks and a Howard 24:1 reduction gearbox. Now known as the Howard Rotavator Company Limited, the firm moved to Norfolk in the 1970s and concentrated more on farm implements and machinery. Production of the reduction box was taken up by County Power Drives, which continued with two versions, 24:1 and 4.6:1.

Many other companies based their drainage machines on Ford tractor skid units, including John Allen of Oxford, Barth and Bruff. The two latter companies also built wheeled drainage machines using County tractors. The McConnel-Thornton-Garnett TG3 trenchless drainer was initially demonstrated on a Roadless 115 four-wheel drive tractor, and Hoes and Mastenbroek made extensive use of Ford industrial engines in their machines.

Most firms involved in the manufacture of front-end loaders built an attachment of some sort to fit the Ford, including Compton, Horndraulic, Mil, Whitlock, and the SkyHi loader built by William Moseley of Feltham. Cameron Gardner designed the Rearloda, which was fitted on the back of the tractor. The Horndraulic loader was sold by Steel Fabricators of Cardiff, manufacturers of the Shawnee industrial loader for the Fordson Major. The company also marketed the Shawnee Poole rear dump trucks fitted to Ford tractors. Whitlock Bros Limited, of Great Yeldham in Essex, also built a number of excavators, loaders and dumpers based on Fordson tractors.

In addition to the various examples of construction machinery built by the companies already mentioned in this book, the Ford tractor was used as the basis of a whole host of different machines used in industrial applications all over the world, all too numerous to mention. Even in Australia, the Melbourne Ford dealers, Malcolm Moore Limited, built a considerable number of conversions on Ford tractors dating back to the 1920s, including rollers, loaders and graders.

RIGHT: A Fordson Super Major equipped with a backhoe and loader built by the American Sherman company.

KEEP CLEAR

Appendices

County Tractors: Models and Specifications

This list contains all the principal tractor models manufactured by County. It has been compiled from various sources of information made available to the author and is intended as an approximate guide only.

The transmissions used in the tractors were often modified or strengthened by County.

– indicates no available information.

Model	Prefix	Built from	To	Engine	Transmission	Comments
Full Track	CFT	Dec 1948	Oct 1951	E27N	E27N	Perkins P6 optional
Full Track	CFT/B	Oct 1951	Approx 1955	E27N	E27N	Perkins P6 optional
Full Track Model 'Z'	CFTZ	Dec 1951	1954	E1A Major	E1A Major	
Full Track Mk II	A	Feb 1954	1958	E1A Major	E1A Major	
Full Track Mk III	E	Feb 1954	1956	E1A Major	E1A Major	Industrial
Mk III Bulldozer	BD	1954	1956	E1A Major	E1A Major	
Mk III Swamp	W	1954	1956	E1A Major	E1A Major	
Mk III Extended Swamp	WL	1954	1956	E1A Major	E1A Major	
Mk III Side-Boom Crane	SB	1954	1956	E1A Major	E1A Major	
Full Track Mk IV	4E	Dec 1956	1959	E1A Major	E1A Major	Industrial
Mk IV Bulldozer	4BD	1956	1959	E1A Major	E1A Major	
Mk IV Swamp	4W	1956	1959	E1A Major	E1A Major	
Mk IV Extended Swamp	4WL	1956	1959	E1A Major	E1A Major	
Mk IV Side-Boom Crane	4SB	1956	1959	E1A Major	E1A Major	
Ploughman	P	Oct 1957	1958	E1A Major	E1A Major	
Ploughman P50	P50	1958	1961	Power Major	Power Major	
CD50	CD50	Dec 1959	1965	Power Major	Power Major	Super Major 1961 -
CD50 Angledozer	CDA50	1959	1965	Power Major	Power Major	Super Major 1961 -
CD50 Side-Boom Crane	CDS50	1959	1965	Power Major	Power Major	Super Major 1961 -
Ploughman P55	P55	Jan 1961	1965	592E	Super Major	
Clydesdale	–	Dec 1961	Approx 1964	Super Major	Super Major	
Four-Drive	FD	1954	1958	E1A Major	E1A Major	
Four-Drive Angledozer	FD/B	1954	1958	E1A Major	E1A Major	
Four-Drive Mk II	4FD	1958	1962	Power Major	Power Major	Super Major 1961 -
Four-Drive Angledozer II	4FD/B	1958	1962	Power Major	Power Major	Super Major 1961 -
Super-4	S4	Aug 1961	Feb 1965	Super Major	Super Major	
Super-6	S6	Oct 1962	Sep 1965	590E	Super Major	
Sea Horse	–	1963	1964	Super Major	Super Major	Amphibious
654	4F	Nov 1964	June 1968	5000	5000	
FC654	4FC	May 1965	June 1968	5000	5000	Forward Control
954	6F	May 1965	June 1967	2703E	5000	
854T	4FT	1967	Approx 1968	5000	5000	CAVTurbocharger
754	4F	May 1968	Nov 1975	5000	5000	Ford Force
FC754	4FC	July 1968	Nov 1975	5000	5000	Forward Control
1004	6F	Dec 1966	Jan 1970	2703E	5000	
1004	6F	Jan 1970	Jan 1971	2703C	5000	
1004	6F	Jan 1971	Feb 1975	2713E	5000	
FC 1004	6FC 8FC	Feb 1967	Jan 1970	2703E	5000	Forward Control
FC1004	6FC 8FC	Jan 1970	Jan 1971	2703C	5000	Forward Control,
FC1004	6FC 8FC	Jan 1971	Nov 1975	2713E	5000	Forward Control
FC1004	26FC	Nov 1975	Dec 1977	2713E	6600	Forward Control
1124	7F	July 1967	March 1970	2704E	5000	
1124	7F	March 1970	Jan 1971	2704C	5000	
1124	7F	Jan 1971	July 1971	2714E	5000	

Model	Prefix	Built from	To	Engine	Transmission	Comments
4000/Four	F4S	Sept 1968	Nov 1975	4000	4000	Unequal Wheel
1164	8F	July 1971	Nov 1975	8000	5000	
1164	28F	Nov 1975	April 1977	8600	7600	
944	3F	Dec 1971	Nov1975	7000	7000	Turbocharged
1254	12F	March 1972	July 1975	8000	8000	
1454	14F	March 1972	Oct 1975	9000	9000	Turbocharged
1454	14F	Oct 1975	Dec 1978	9600	9600	Turbocharged
764	24F	Oct 1975	Jan 1978	6600	6600	
964	23F	Oct 1975	Dec 1977	7600	7600	Turbocharged
4600-Four	F24S	Oct 1975	Oct 1981	4600	4600	Unequal Wheel
6600-Four	F6S(F26S)	Dec 1975	Oct 1981	6600	6600	Unequal Wheel
7600-Four	F7S(F27S)	Dec 1975	Oct 1981	7600	7600	Unequal Wheel
6700-Four	F32S	Feb 1978	Oct 1981	6600	6600	Unequal Wheel
7700-Four	F34S	Feb 1978	Oct 1981	7600	7600	Unequal Wheel
774	30F	Feb 1978	Approx 1990	6600	6600	6610 1982-
974	31F	Jan 1978	Approx 1990	7600	7600	7610 1982-
1174	29F	April 1977	July 1979	8700	7600	
FC 1174	29 FC	Dec 1977	Feb 1980	8700	7600	Forward Control
1474	37F	Dec 1978	July 1979	9700	9700	Turbocharged
1474	29F	July 1979	1981	TW20	TW20	Turbocharged
1474	29F	1981	Approx 1987	TW25	TW25	Turbocharged
1184 TW	42F	July 1979	1981	TW10	7600	
1184 TW	42F	1981	Approx 1990	TW15	7600	
FC 1184 TW	43F	Feb 1980	1981	TW10	7600	Forward Control
FC 1184 TW	43F	1981	Approx 1987	TW15	7600	Forward Control
1884	–	Dec 1980	1981	TW30	TW30	Turbo/Intercooled
1884	–	1981	Approx 1987	TW35	TW35	Turbo/Intercooled
6610-Four E	–	Oct 1981	Feb 1983	6610	6610	Unequal Wheel
7610-Four E	–	Oct 1981	Feb 1983	7610	7610	Unequal Wheel
764	–	1978	Approx 1990	As 774	8-Speed	Export
964	–	1978	Approx 1990	As 974	8-Speed	Export
1164 TW	–	1979	Approx 1990	As 1184 TW	8-Speed	Export
1464	–	1978	Approx 1987	As 1474	8-Speed	Export
1864	–	1980	Approx 1987	As 1884	8-Speed	Export
764P	–	1982	–	A4.248	8-Speed	Export-Perkins
1164P	–	1982	–	A6.354	8-Speed	Export-Perkins
HSH 140	–	1994		8340	8-Speed	High-Speed 2wd.
764 (40)	–	April 1995		6640	8-Speed	Export
1164 (40)	–	April 1995		8340	8-Speed	Export
1184 (40)	–	April 1995		8340	8-Speed	Export/Cab
Highway Dexta	–	1960	1964	Dexta	Dexta	Highway/Industrial
Hi-Drive	–	August 1958	1964	Power Major	Power Major	Super Major 1961-
652H	–	1964	1968	5000	5000	High Clear 2wd.
752H	–	June 1968	1975	5000	5000	High Clear 2wd.
762H	–	1975		6600	6600	High Clear 2wd.
962H	–	1975		7600	7600	High Clear 2wd.
1162	–	1980	Approx 1987	TW10	8-Speed	Export 2wd.
1162P	–	1982	–	A6.354	8-Speed	Export 2wd.Perkins
1162H	–	1980	Approx 1987	TW10	8-Speed	Export High Clear

Doe Triple D & 130 Tractors - Numbers sold

	Doe Triple D							Doe 130	
County (United Kingdom)	**1958**	**1959**	**1960**	**1961**	**1962**	**1963**	**1964**	**1965**	**Total**
Bedfordshire	–	1	–	1	–	2	–	–	4
Berkshire	–	–	–	–	–	2	–	2	4
Cambridgeshire	–	–	1	2	1	2	1	4	11
Derbyshire	–	–	–	–	–	–	1	–	1
Dorset	–	–	–	1	–	–	–	–	1
Essex	1	5	11	23	17	37	10	19	123
Hampshire	–	–	–	–	–	1	2	8	11
Hertfordshire	–	2	3	3	5	4	1	4	22
Huntingdonshire	–	–	–	–	–	1	–	1	2
Kent	–	–	2	1	–	–	1	–	4
Lancashire	–	–	–	–	–	–	1	–	1
Leicestershire	–	–	–	1	1	1	–	1	4
Lincolnshire	–	–	2	3	12	6	4	4	31
Middlesex	–	–	–	–	–	1	1	–	2
Norfolk	–	–	–	–	4	5	1	2	12
Northamptonshire	–	–	–	–	–	1	–	1	2
Nottinghamshire	–	1	–	–	1	–	1	–	3
Oxfordshire	–	–	–	–	–	3	3	–	6
Scotland	–	–	–	–	–	1	–	1	2
Somerset	–	–	–	–	–	1	–	–	1
Suffolk	2	1	4	12	10	5	8	6	48
Sussex	–	–	–	–	–	–	1	1	2
Warwickshire	–	–	–	1	–	2	2	1	6
Yorkshire	–	–	–	1	2	1	2	4	10
Total (UK)	3	10	23	49	53	76	40	59	313
Export (Countries)									
Australia	–	–	–	–	–	–	–	1	1
Austria	–	–	–	–	–	–	1	–	1
Denmark	–	–	–	–	–	–	1	2	3
Germany	–	–	–	1	1	3	–	–	5
Ireland	–	–	1	–	–	–	–	–	1
Nigeria	–	–	–	–	1	–	–	8	9
South Africa (Cape Town)	–	–	–	–	1	1	–	–	2
Southern Rhodesia	–	–	–	–	–	–	1	–	1
Spain	–	–	–	–	1	–	–	–	1
Sweden	–	1	2	3	–	–	–	3	9
Uruquay	–	–	–	1	–	–	–	–	1
USA	–	–	–	1	3	–	–	–	4
USSR	–	2	–	–	–	–	1	–	3
Total (Export)	–	3	3	6	7	5	4	14	42
Overall Total	**3**	**13**	**26**	**55**	**60**	**81**	**44**	**73**	**355**

NB: These numbers may not be complete as some records were destroyed by fire.

Details of Present Day Companies

Agrosave, The Sidings, Abbots Salford Road,
Harvington, Evesham.
Roadless tractor spares.

Aveling Barford (Machines) plc, Invicta Works, Grantham, Lincolnshire,
NG31 6JE.
Sanderson farm handlers and construction equipment.

J. C. Bamford Excavators Limited, Rocester,
Staffordshire, ST14 5JP.
JCB backhoe loaders, wheeled loaders, hydraulic excavators, high-speed tractors and other handling, earthmoving and construction equipment.

County Tractors Limited, Ludlow Road, Knighton, Powys, LD7 1LP.
County tractors and 4x4 Ford Transits.

Ernest Doe & Sons Limited, Ulting, Maldon, Essex, CM9 6QH.
Agricultural and construction machinery dealers for a wide range of equipment including New Holland Ford.

Dowdeswell Engineering Company Limited, Blue Lias Works, Stockton,
Rugby, Warwickshire, CV23 8LD.
Dowdeswell ploughs and tillage equipment.

Lloyd Loaders (M.H.) Limited, Mytholm Works, Hipperholme, Halifax,
West Yorkshire, HX3 8PJ.
New, used and reconditioned Muir-Hill loading shovels and tractors.

Matbro Limited, London Road, Tetbury, Gloucestershire, Gl8 8JD.
Matbro telescopic handlers and loading shovels.

New Holland Limited, Cranes Farm Road, Basildon,
Essex, SS14 3AD.
Farm and construction equipment including Ford and Fiatagri tractors, and New Holland combines.

A. T. Osborne Limited, Ower, Romsey, Hampshire, S051 6ZL
Forestry and agricultural engineers, and specialists in County tractors.

Thama Holdings
Dale Street, Bilston, West Midlands WV14 7JU.
Spares for County tractors and Bamford machinery.

Jas. P. Wilson, Industrial Site, Coast Road, Dalbeattie, DG5 4QU.
County tractor specialists and dealers in forestry, drilling and construction equipment.

Bibliography

JOURNALS & PERIODICALS

Arable Farmer
Commercial Motor
Contractors Plant Review
County News
Doe News
Farm
Farm Contractor
Farm Mechanisation
Farmer and Stockbreeder
Farmers Weekly
Ford Dealer and Service Field
Hertfordshire Countryside
Implement and Farm Machinery Review
Landmark
Mechanical Handling
Muck Shifter
Power Farming
Roadless News

BOOKS & OTHER PUBLICATIONS

Aveling Barford Limited, **A Hundred Years of Road Rollers**, Oakwood Press 1965.
Briggs, Ernest L., **Muir-Hill Album**, Allan T. Condie 1982.
Baldwin, Nick, **Vintage Tractor Album**, Marshall Harris & Baldwin Ltd. 1980.
Batchelor, John H., and Macksey, Kenneth, **Tank**, Macdonald & Jane's 1970.
Cavill, R. A., **A Survey of Four Wheel Drive Applications and Four Wheel Drive Power Transmissions in Agriculture**, (Thesis) Essex Institute of Agriculture 1964.
Ford Tractor Division, **Highlights of Ford in Agriculture**, Ford Motor Company 1962
Hawkins, J. C., **The Mechanical Equipment of Farms**, Farmer & Stockbreeder 1949.
Hine, H. J., **Farm Mechanics Notebook,** Farmer & Stockbreeder 1948.
Ingram, Arthur, **Ford Trucks and Transport since 1945**, Motor Racing Publications 1978.
Olyslager Organisation, **Earthmoving Vehicles**, Frederick Warne & Co. 1972.
Williams, Michael, **Ford & Fordson Tractors**, Farming Press, 1985.

Various manufacturers' publications and sales catalogues.

Index

Roman numbers refer to text, bold to illustrations

K

L

M

N

O

P

R

S

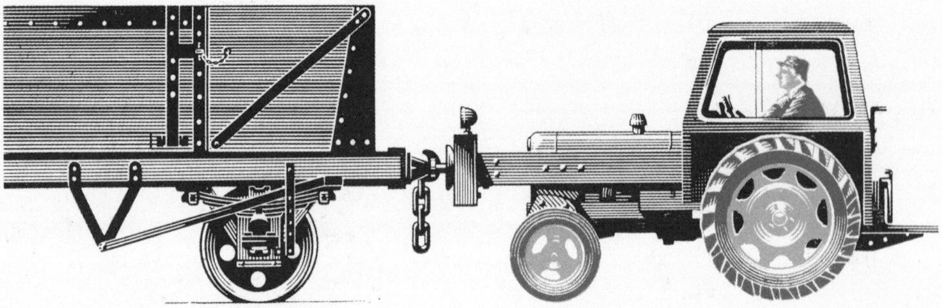

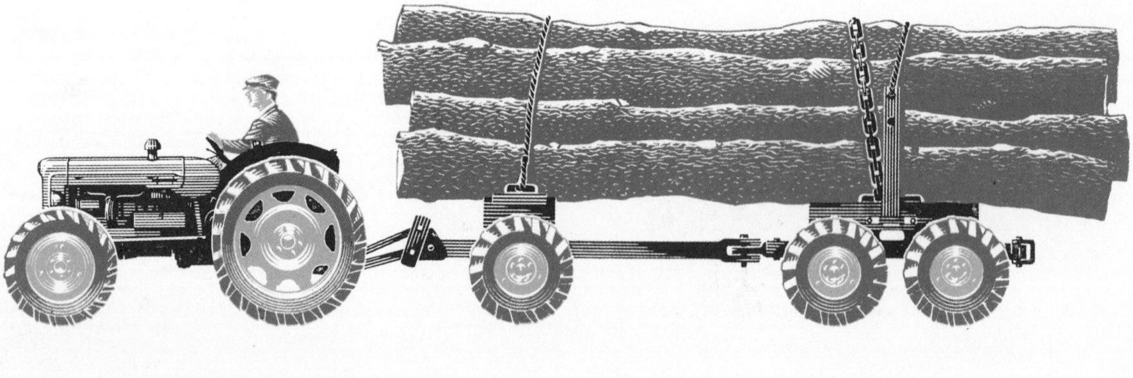

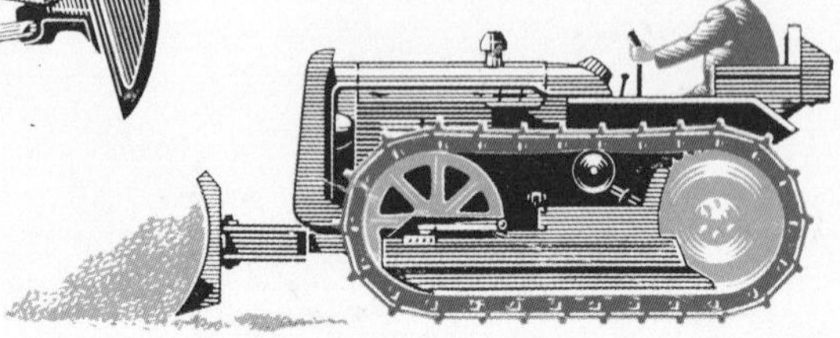